Manufacturing for the 21st Century 2014

Topical Conference at the 2014 AIChE Spring Meeting and 10th Global Congress on Process Safety

New Orleans, Louisiana, USA
30 March – 3 April 2014

ISBN: 978-1-63439-060-6

Printed from e-media with permission by:

Curran Associates, Inc.
57 Morehouse Lane
Red Hook, NY 12571

CURRAN ASSOCIATES INC.
proceedings
.com

Some format issues inherent in the e-media version may also appear in this print version.

Printed by Curran Associates, Inc. (2014)

For permission requests, please contact AIChE
at the address below.

AIChE
3 Park Avenue
New York, NY 10016-5991

Phone: (203) 702-7660
Fax: (203) 775-5177

www.aiche.org

Additional copies of this publication are available from:

Curran Associates, Inc.
57 Morehouse Lane
Red Hook, NY 12571 USA
Phone: 845-758-0400
Fax: 845-758-2634
Email: curran@proceedings.com
Web: www.proceedings.com

TABLE OF CONTENTS

354012 Possible vs. Practical: Engineers Must Lead the Development of Practical Technologies

Abstract:

Chemical Engineering is need today more than ever. There is a strong societal desire for sources of energy and feedstocks that can sustainably support a high standard of living for all. Thermodynamic, material and economic challenges stand in the way. Most of the general public does not have the basic understanding of the energy flows, chemical transformations, and the scale of the systems that support our current way of life. These concepts are fundamental to chemical engineering. Chemical engineers should be on the vanguard of separating hype from truly practical options yet occationally we let society down. This talk will present several examples where the fundamentals of chemical engineering can be used to access the practicality of proposed solutions to energy and water issues, as well as explain why the venture capital model struggles in the materials market.

352696 Hydrocarbons from Biogas Via Biogas Reforming, Syngas Cleaning and Catalytic Conversion

Abstract:

This research work focused on taking advantage of biogas as a cleaner, sustainable and renewable feedstock to produce liquid hydrocarbons. In this approach, biogas is first reformed to high quality synthesis gas (H_2 + CO) followed by Fischer-Tropsch synthesis (FTS). The Sustainable Energy Research Center (SERC) at Mississippi State University has developed and tested a continuous process over a pilot plant system to demonstrate the technology of biomass to liquid (BTL) fuels. The process includes biomass gasification, syngas conditioning and cleaning, and FTS steps. Currently, the typical gaseous product components from the gasifier are 20% H_2, 19% CO, 12% CO_2, and 47% N_2 if using wood chips as the feedstock. The N_2 and CO_2 contents are too high, the total syngas (H_2 + CO) content is only about 40%, and H_2/CO ratio of the bio-syngas is less than 1. The biomass-derived product gas is not ideal for the hydrocarbon synthesis. Exceptional product gas is produced from biogas (Table 1), it contains high concentrated syngas (CO+ H_2 > 90%), low impurities and CO_2, as well as N_2,H_2/CO ratio is between 1 and 2 which makes it perfect production of liquid hydrocarbon by FTS. The liquid products generated by FTS can be readily upgraded to transportation fuels since they are primarily composed of hydrocarbons initially.

Table 1 Product gas composition for simultaneous dry and partial oxidation (using air) of biogas.

Temperature	H_2 (%)	CO (%)	N_2 (%)	CH_4 (%)	CO_2 (%)	H_2/CO

(°C)						
750	48.96	40.82	6.33	1.75	2.01	1.2
775	49.45	41.22	6.27	1.3	1.74	1.2
800	50.06	42.2	6.05	0.96	1.12	1.19
825	50.57	42.7	6.03	0.57	0.78	1.18
850	50.75	43.11	5.93	0.4	0.25	1.18

351056 Research for an Energy Future

Abstract:

Global demand for electrical energy is increasing even as environmental concerns are changing the ways in which we produce it. Southern Company is a leading US producer of electricity and is dedicated to developing the latest technologies of pollution control, energy production, and energy efficiency, and bringing them to economic feasibility. This talk will highlight efforts in all these areas, including Southern Company's major research projects, research centers, and commercial operations. The talk will primarily focus on Southern Company's roles in developing pollution control technology, including acid gas scrubbers, selective catalytic reduction of NOx, and mercury control (including the Mercury Research Center and the High-Sulfur Pilot Baghouse Research Center). CO_2 capture will be discussed including a 25-MW demonstration plant and ongoing work at the National Carbon Capture Center, located at Alabama Power's Gaston Plant. Changing energy production methods including commercial operations of coal gasification, wind, and solar will also be discussed, as well as the new Solar Power Research Center. Efforts to develop technologies which conserve water in power production will be highlighted by a discussion of the Water Research Center located at Georgia Power's Bowen Plant. Finally, ongoing research to improve energy efficiency for our customers will be highlighted, as well as electric transportation research which would have major environmental benefits.

Sustainability and Environmental Impacts of Biofuel

Shijie Liu

Department of Paper and Bioprocess Engineering
SUNY College of Environmental Science and Forestry

1

Imagine a renewable world

2

Outline

- The Hype of Biomass
- What is Sustainability?
- Biomass Crop
- Ethanol: Good or Bad
- Ethanol: a Platform chemical
- Ethanol: Biomass conversion efficiency
- Conclusion

3

The Hype in Biomass

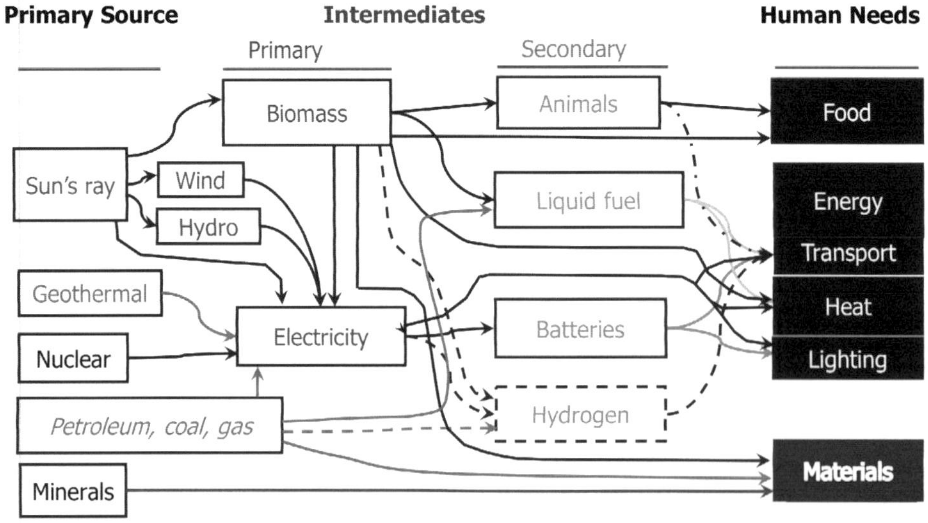

4

Where do we get our chemicals today?

- Petroleum, coal and natural gas
  - Dominant source of energy for human use
  - Materials use - plastics
- Biomass
  - Food
  - Heating use on the steep declining slope
  - Materials – dwellings, furniture
- Solar energy
  - Limited use in heating and lighting

5

What is Renewability?

- Merriam-Webster:
  - 1. Renewable – capable of being renewed
  - 2. Renewable – capable of being replaced by natural ecological cycles or sound management practices
- Point on a complete cycle

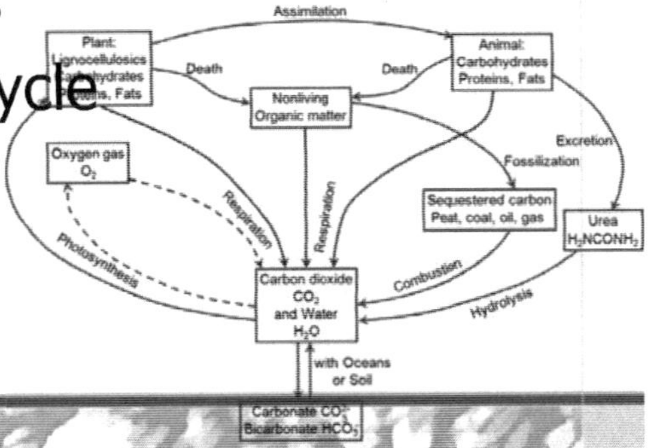

What is Sustainability?

- Merriam-Webster:
  - 1. Sustainable – capable of being sustained
  - 2. Sustainable – of, relating to, or being a method of harvesting or using a resource so that the resource is not depleted or permanently damaged.
- Non-trivial Steady State

  (S. Liu, Bioprocess Engineering: Kinetics, Biosystems, Sustainability, and Reactor Design, Elsevier, 2012)

Biomass

- Agricultural crop
  - Cereal grain; Tubers
  - Straws
  - Grass
- Forest biomass
  - Trees
- Long rotation crops
  - Fruits, nuts
  - "Short-rotation" woody biomass

8

Is Forest Biomass Renewable?

- Manomet Center for Conservation Sciences report "Biomass Sustainability and Carbon Policy Study" also know as the Massachusetts Biomass Report (June 2010)
  - Woody biomass utilization sending more carbon into the atmosphere than fossil
  - Equal setting of forest biomass use with fossil use will occur after 32 years IF ...
  - Forest biomass carbon neutrality will not occur after 90 years

9

Massachusetts' report

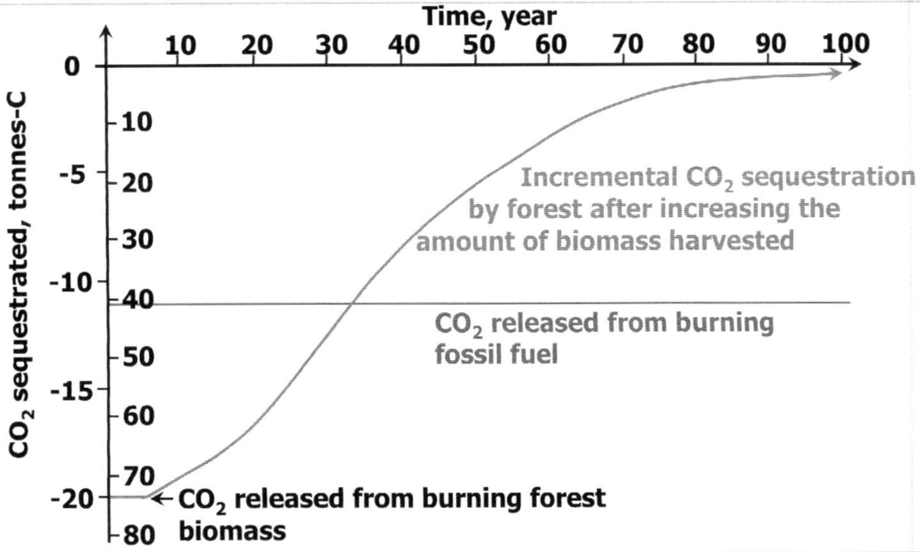

10

Biomass: Good or Bad?

- Fossil source
  - Energy dense: low CO_2 emission
- Biomass, CO_2 is CO_2
  - Less energy content (already oxygenated) - bulky
  - Not accounting: Forest disturbance – only half storage grows back when converting an untouched forest to a managed forest and equal amount underground in jeopardy
  - More CO_2 in atmosphere at sustainability
  - Hydrocarbon emission, e.g. isosprene

Sustainability – Biomass Production

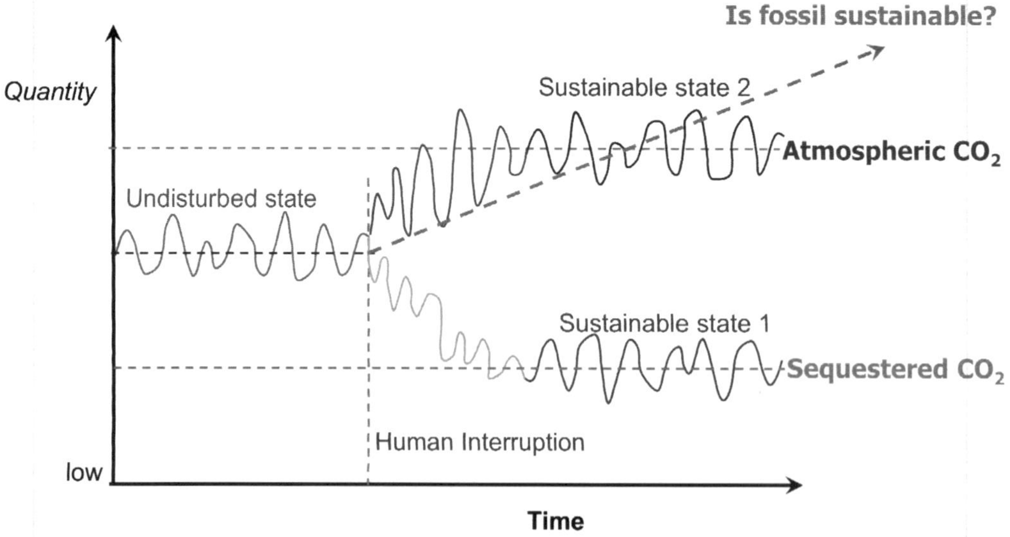

What is the Magic Biomass (Crop)?

Table 1. Recycle times and productivities of chemical/energy feedstock (Liu, 2010).

Feed stock	Recycle time	Standing Biomass tons/ha	Biomass Production tons/(ha·year)
Algae	1 month	0.9	11.25
Agricultural crops	3 month - 1 year	4.5	2.93
Temperate grasses	1 year	7.2	2.70
Savanna	1 year	18.0	4.05
Shrubs	1 – 5 years	27.0	3.15
Tropical forest	5 – 25 years	202.5	9.90
Trop. season forest	5 – 25 years	157.5	7.20
Boreal forest	25 – 80 years	90.0	3.60
Temperate deciduous	10 – 50 years	135.0	5.40
Temperate evergreen	10 – 80 years	157.5	5.85
Oil, gas and coal	280 million years	(38.4×10^{21} J)	(0)

Standing Biomass (~ X_∞)

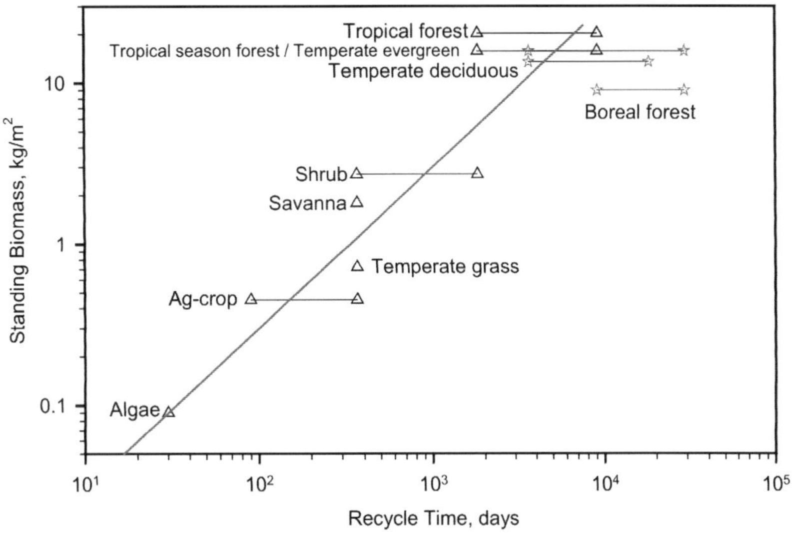

14

Biomass crop selection

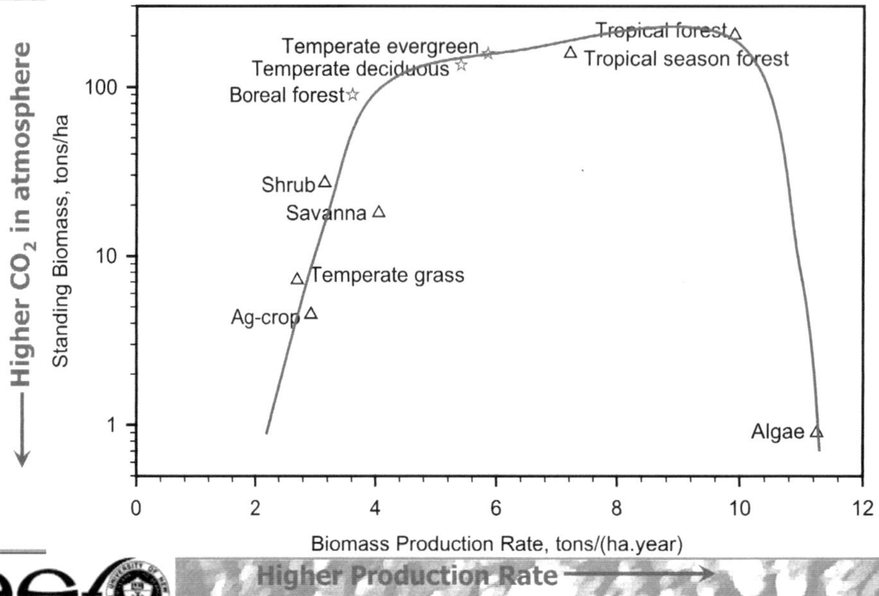

Biomass Crop Selection

- Productivity: rate of supply
- Standing biomass: CO_2 sequestration
- Water, Soil condition, Terrain
- Neighboring community
- Food needs, why "corn" is a better biomass
- Production cost
- There is no magic crop

Integrated Biorefinery:

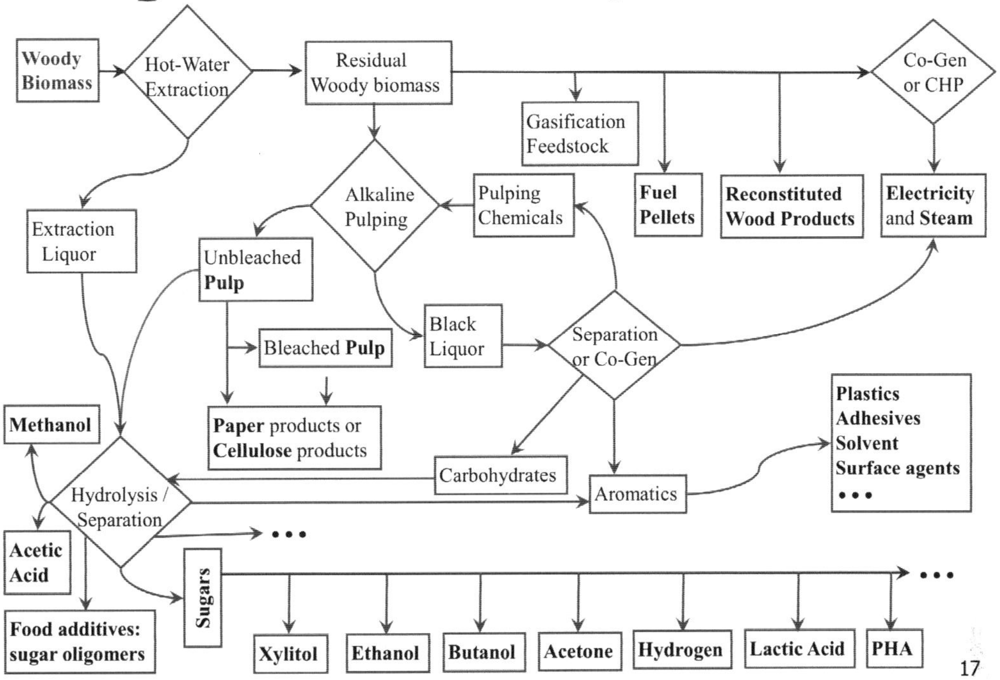

17

Ethanol: Good or Bad

- Gasoline additive: E10, going for E15
  - Commercial
  - Infrastructure established, and production facility design mature for new ones
  - Blending at distribution centers
  - Transportation
- Cereal grain use
- Physical and chemical properties

Ethanol: Platform chemical

- Jet fuel
- Bioplastics or bioploymers
- Hydrocarbons

$$CH_3CH_2OH \longrightarrow CH_2{=}CH_2 + H_2O$$

$$2CH_3CH_2OH \longrightarrow CH_2{=}CH\text{-}CH{=}CH_2 + H_2O + 2H_2$$

Wang, Y. and Liu, S. J. Bioproc. Eng. Biorefinery, 1(1), 33-43.

Ethanol: Biomass conversion

- Theoretical yield: 51% on sugars
  - Anaerobic fermentation, can tolerate O_2
  - Fermentation to high concentrations
- As compared with
- Butanol – low yield, low concentration
- PHA – low yield, high recovery cost

Fermentation: Wood extract Hydrolysate

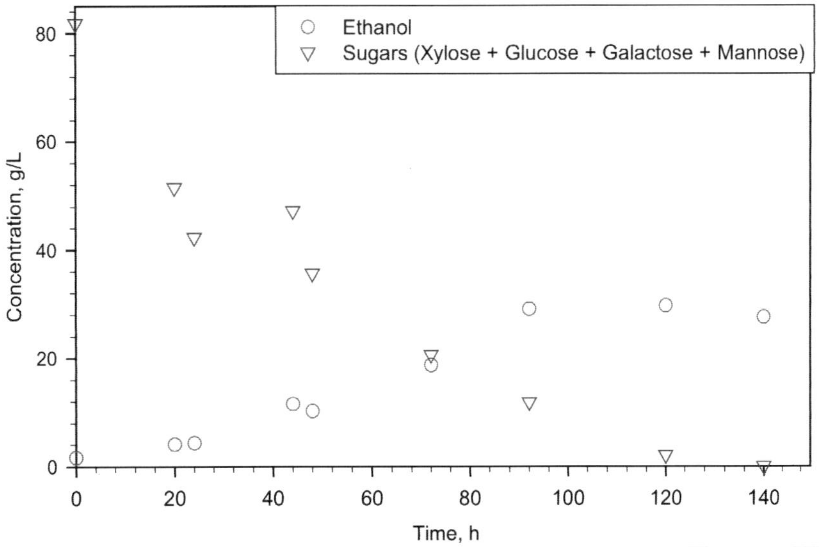

Butanol fermentation

Glucose in feed: 45 g/L

Acetone	5.2	g/L
Ethanol	1.8	g/L
Butanol	9.8	g/L
Total	16.8	g/L

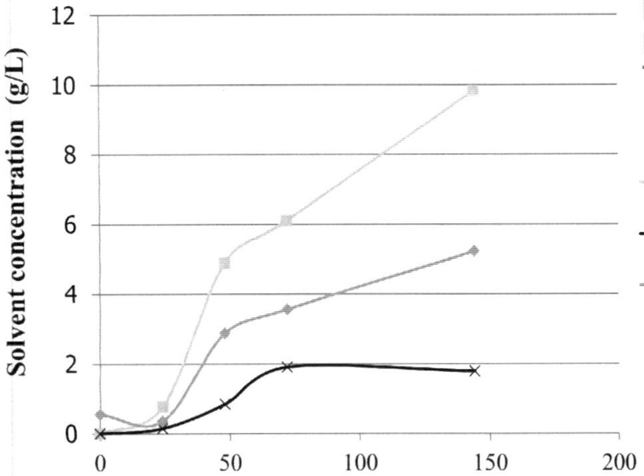

Butanol fermentation

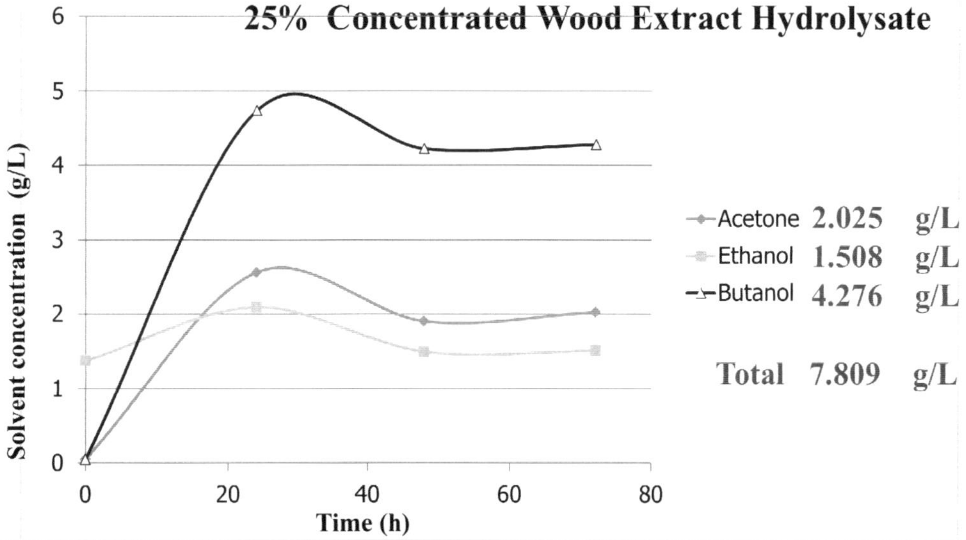

Conclusions

- Future depends on biomass, despite a hard choice to make: CO_2 is CO_2
- Biomass is both renewable and sustainable
- Ethanol is a good choice for continuing developments in renewable chemical and energy sectors
- Ethanol is a platform chemical, from which products know today from petroleum can be produced

24

Thanks!

25

352830 Air Permitting of a Major Capital Project at a Chemical Plant in Louisiana -

Abstract:

ABSTRACT for 2014 AIChE Spring Meeting – New Orleans iE Sprin Modn

Authors:

Steve David, Senior Manager of Air Strategy, Mosaic Corporation

Peter H. Anderson, Principal, Geosyntec Consultants

Air Permitting of a Major Capital Project at a Chemical Plant in Louisiana - As part of preparing a prevention of significant deterioration (PSD) and Part 70 (Title V) permit application for a proposed $1 billion expansion at a chemical plant a number of technical analyses were completed to demonstrate that the Project would meet federal and state air pollution control requirements. Included as part of this effort, was implementing a comprehensive permitting strategy to evaluate control technology options to minimize regulatory impacts and executing a plan to reduce critical non-technical risks to support an expedited permitting process. This paper will provide insights on how to permit a complex expansion of a chemical plant and cover the challenging topics of determining best available control technology (BACT) for greenhouse gases (GHG) and CO_2 emissions and carrying-out air dispersion modeling to demonstrate compliance with national ambient air quality standards (NAAQS) and state toxic air pollutant ambient air standards.

Theoretical Screening of Mixed Solid Sorbent for Applications to CO_2 Capture Technology

Yuhua Duan[1]
National Energy Technology Laboratory, United States Department of Energy, Pittsburgh, Pennsylvania 15236, USA

Abstract

Since current technologies for capturing CO_2 to fight global climate change are still too energy intensive, there is a critical need for development of new materials that can capture CO_2 reversibly with acceptable energy costs. Accordingly, solid sorbents have been proposed to be used for CO_2 capture applications through a reversible chemical transformation. By combining thermodynamic database mining with first principles density functional theory and phonon lattice dynamics calculations, a theoretical screening methodology to identify the most promising CO_2 sorbent candidates from the vast array of possible solid materials has been proposed and validated. The calculated thermodynamic properties of different classes of solid materials versus temperature and pressure changes were further used to evaluate the equilibrium properties for the CO_2 adsorption/desorption cycles. According to the requirements imposed by the pre- and post- combustion technologies and based on our calculated thermodynamic properties for the CO_2 capture reactions by the solids of interest, we were able to screen only those solid materials for which lower capture energy costs are expected at the desired pressure and temperature conditions. Only those selected CO_2 sorbent candidates were further considered for experimental validations. The *ab initio* thermodynamic technique has the advantage of identifying thermodynamic properties of CO_2 capture reactions without any experimental input beyond crystallographic structural information of the solid phases involved. Such methodology not only can be used to search for good candidates from existing database of solid materials, but also can provide some guidelines for synthesis new materials. In this presentation, we apply our screening methodology to mixing solid systems to adjust the turnover temperature to help on developing CO_2 capture Technologies.

I. Introduction

Carbon dioxide is one of the major combustion products which once released into the air can contribute to global climate change.[1-4] One approach to solve such environmental problems is to capture and sequester the CO_2. Current technologies for capturing CO_2 including solvent-based (amines) and CaO-based materials are still too energy intensive. Hence, there is critical need for development of new materials that can capture and release CO_2 reversibly with acceptable energy costs. In particular, solid oxide sorbent materials have been proposed for capturing CO_2 through a reversible chemical transformation leading primarily to formation of carbonate products. Solid sorbents containing alkali and alkaline earth metals have been reported in several previous studies

[1] Tel. 412-386-5771, Fax 412-386-5920, E-mail: yuhua.duan@netl.doe.gov

to be promising candidates for CO_2 sorbent applications due to their high CO_2 absorption capacity at moderate working temperatures.[5-8]

During past few years we developed a theoretical methodology to identify promising solid sorbent candidates for CO_2 capture by combining thermodynamic database searching with *ab initio* thermodynamics obtained based on first-principles density functional theory (DFT) and lattice phonon dynamics.[5-7, 9-13] As shown in Fig.1, the primary outcome of our screening scheme is a list of promising CO_2 sorbents with optimal energy usage.

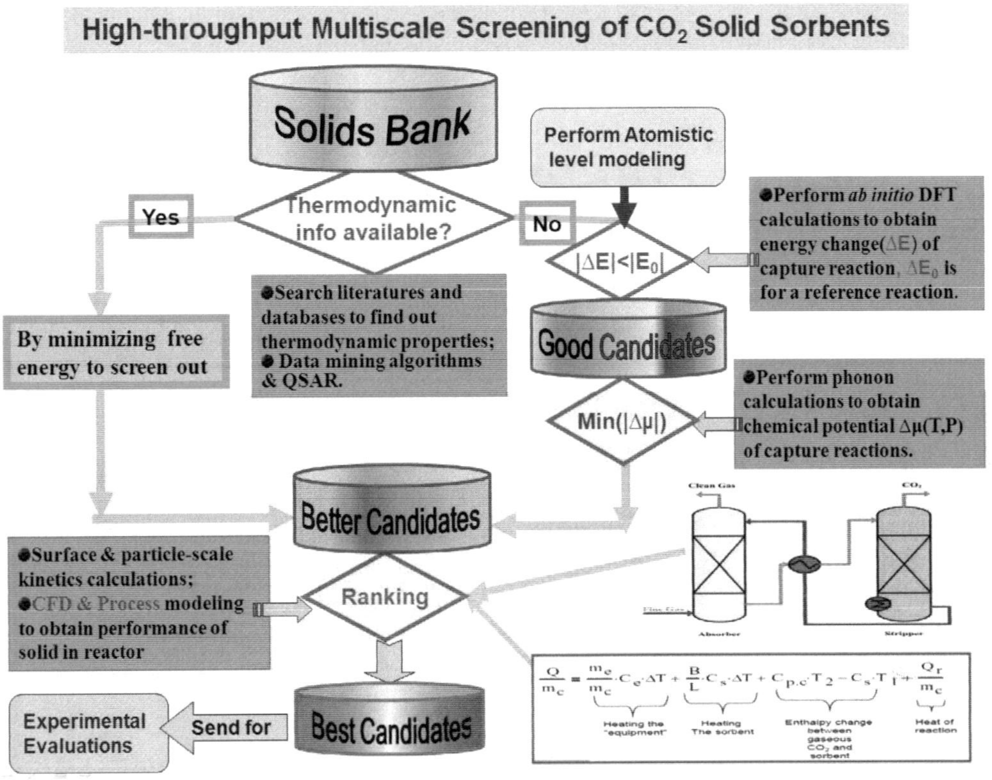

Fig. 1. Schematic of our screening methodology.

At a given CO_2 pressure, the turnover temperature (T_t) of an individual solid capture CO_2 reaction is fixed. Such T_t may be outside the operating temperature range (ΔT_o) for a particularly capture technology. In order to adjust T_t to fit the practical ΔT_o, in this study, we demonstrate that by mixing different types of solids it's possible to shift T_t to the range of practical operating conditions.

II. Calculation Methods for Mixed Solid Sorbents

The complete description of the computational methodology together with relevant applications can be found in our previous publications.[5-7, 9-16] The CO_2 capture reactions of solids can be expressed generically in the form (for convenient description, we normalized the reaction to 1 mole of CO_2)

$$\sum_{Ri} n_{Ri} Solid\_R_i + CO_2 \leftrightarrow \sum_{Pj} n_{Pj} Solid\_P_j \qquad (a)$$

where n_{Ri}, n_{Pj} are the numbers of moles of reactants (R_i) and products (P_j) involved in the capture reactions. We treat the gas phase CO_2 as an ideal gas. By assuming that the difference between the chemical potentials ($\Delta\mu^o$) of the solid phases of reactants (R_i) and products (P_j) can be approximated by the difference in their total energies (ΔE^{DFT}), obtained directly from DFT calculations, and the vibrational free energy of the phonons and by ignoring the PV contribution terms for solids, the variation of the chemical potential ($\Delta\mu$) for reaction (a) with temperature and pressure can be written as[5-7, 9-16]

$$\Delta\mu(T,P) = \Delta\mu^0(T) - RT \ln \frac{P_{CO_2}}{P_0} \qquad (1)$$

where

$$\Delta\mu^0(T) \approx \Delta E^{DFT} + \Delta E_{ZP} + \Delta F^{PH}(T) - G^0_{CO_2}(T) \qquad (2)$$

Here, ΔE^{DFT} is the DFT energy difference between the reactants and products of the reaction (a), ΔE_{ZP} is the zero point energy difference between the reactants and products and can be obtained directly from phonon calculations. ΔF^{PH} is the phonon free energy change excluding zero-point energy (which is already counted into the ΔE_{ZP} term) between the solids of products and reactants. P_{CO_2} is the partial pressure of CO_2 in the gas phase and P_0 is the standard state reference pressure taken to be 1 bar. The heat of reaction ($\Delta H^{cal}(T)$) can be evaluated through the following equation

$$\Delta H^{cal}(T) = \Delta\mu^0(T) + T[\Delta S_{PH}(T) - S_{CO_2}(T)] \qquad (3)$$

where $\Delta S_{PH}(T)$ is the difference of entropies between product solids and reactant solids. The free energy of CO_2 ($G^0_{CO_2}$) can be obtained from standard statistical mechanics,[6, 7, 13] and its entropy (S_{CO_2}) can be found in the empirical thermodynamic databases.[17]

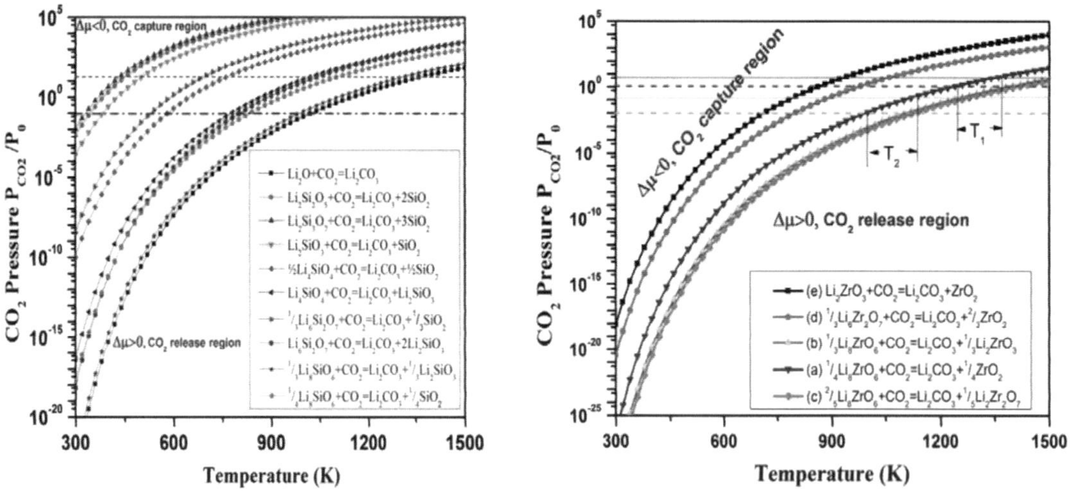

(a) Li_2O/SiO_2 (b) Li_2O/ZrO_2

Fig. 2. Contour plots of the calculated chemical potential ($\Delta\mu$) versus temperature and the CO_2 pressure (*P* plotted in logarithmic scale) for the CO_2 capture reactions. Only $\Delta\mu$=0 curve is shown explicitly. For each reaction, above its $\Delta\mu$=0 curve, their $\Delta\mu$<0, which means the sorbents absorb CO_2 and the reaction goes forward, whereas below the $\Delta\mu$=0 curve, their $\Delta\mu$>0, which indicates CO_2 start to be released and reaction goes backward with regeneration of the sorbents.

III. Results and Discussions

When we mix two solids *A* and *B* to form a new sorbent *C*, the turnover temperature of the newly resulted system (T_C) is located between those of *A* and *B* (T_A, T_B). Here it was assumed that *A* is a strong CO_2 sorbent while *B* is a weak CO_2 sorbent and $T_A>T_B$. Also, we assumed that the desired operating temperature T_O is between T_A and T_B ($T_A>T_O>T_B$). Now, depending on the properties of *A* and *B*, we have typically three scenarios to synthesize the mixing sorbent *C*:

3.1. $T_A>>T_B$ and the *A* component is the key part to capture CO_2.

An example of this case is represented by Li_2O. This is a very strong CO_2 sorbent which forms Li_2CO_3. However, its regeneration from Li_2CO_3 only can occur at very high temperature (T_A). In order to move its T_A to lower temperatures, one can mix some weak CO_2 sorbents (such as SiO_2, ZrO_2). Fig. 2 shows the relationship of chemical potential, P_{CO2}, and T of the CO_2 capture reactions by the mixed Li_2O/SiO_2 and Li_2O/ZrO_2 solids with different mixing ratios. Fig. 3 shows the turnover temperature and the CO_2 capture capacity of Li_2O/SiO_2 mixture versus the ratio of Li_2O/SiO_2.[5-7, 9, 15, 18-20]

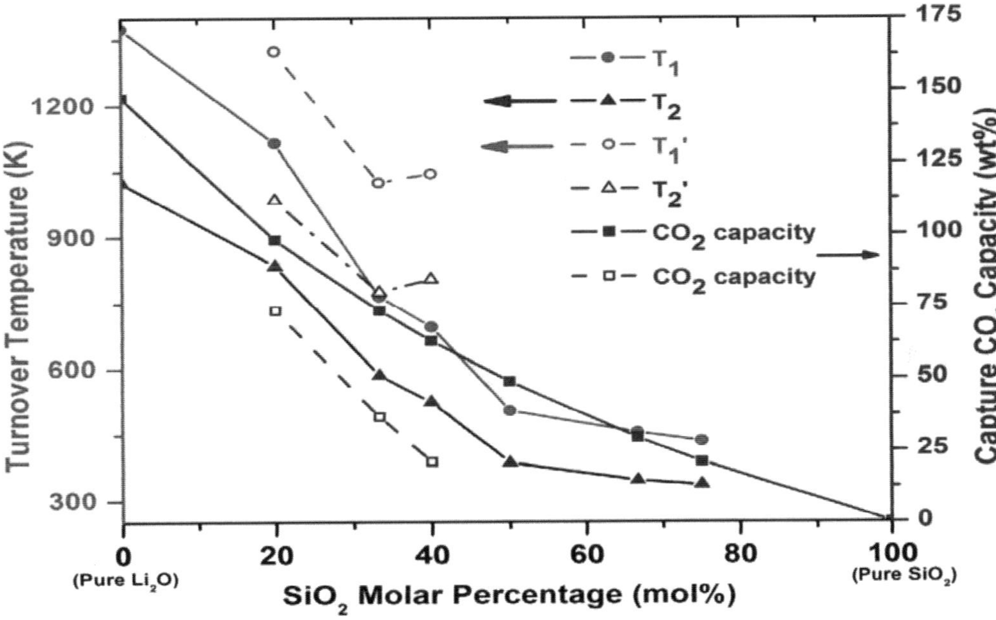

Fig. 3 The dependence of the turnover temperatures defined in the text and of CO_2 capture capacity on molar percentage of SiO_2 in the silicates for which calculations are reported here. T_1 and T_1' are the turnover temperatures under pre-combustion conditions with CO_2 partial pressure at 20 bars, while T_2 and T_2' are the turnover temperatures under post-combustion conditions with CO_2 partial pressure at 0.1 bar. The solid lines indicate to convert lithium silicates to SiO_2 and Li_2CO_3 (T_1, T_2). For those Li_2O-rich lithium silicates (Li_8SiO_6, Li_4SiO_4, $Li_6Si_2O_7$) capturing CO_2, the data shown in dash lines indicate convert them to Li_2SiO_3 and Li_2CO_3 (T_1', T_2'). The corresponding CO_2 capture capacities are plotted with open blue squares.

From Figs. 2 and 3, one can see that after mixing Li_2O and SiO_2 (or ZrO_2) with different Li_2O/SiO_2 (or Li_2O/ZrO_2) ratios, the T_C of the newly formed *C* compound (silicate or zirconate) is lower than T_A of pure Li_2O and could be close to the ΔT_o range to fit the practical needs.

3.2 T_A>>T_B and B component is the key part to capture CO_2

In this case, since T_B is lower than T_O, by mixing *A* into *B* will increase the turnover temperature T_C of the *C* solid to values closer to T_o. For example, pure MgO has a very high theoretical CO_2 capture capacity. However, its turnover temperature (250 °C) is lower than the required temperature range of 300-470 °C used in warm gas clean up technology and its practical CO_2 capacity is very low, and therefore, it cannot be used directly as a CO_2 sorbent in this technology.[21-23] As shown in Fig. 4, by mixing alkali metal oxides M_2O (M=Na, K, Cs, Ca) or carbonates (M_2CO_3) into MgO, the corresponding newly formed mixing systems have higher turnover temperatures making them useful as CO_2 sorbents through the reaction MgO + CO_2 + M_2CO_3 = $M_2Mg(CO_3)_2$.[21, 23]

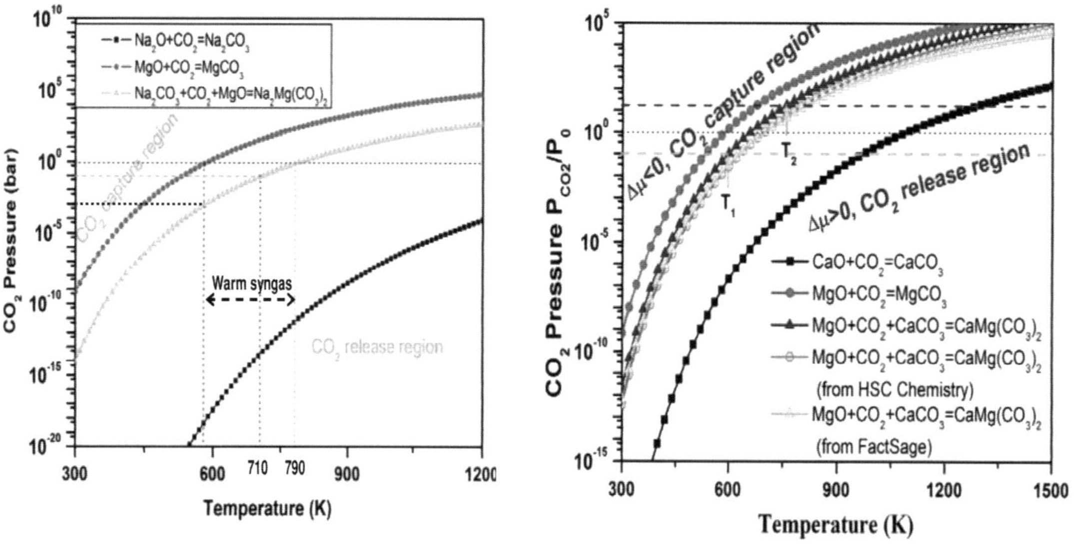

(a) MgO/Na_2CO_3 (Na_2O) (b) $MgO/CaCO_3$ (CaO)

Fig. 4. Plots of the calculated chemical potentials versus CO_2 pressures and temperatures for the CO_2 capture reaction by MgO to form double salts.

3.3 T_A and T_B are close to each other and both A and B are active for CO_2 capture

In this case, both *A* and *B* components are active to capture CO_2, and the CO_2 capacity of the mixture is the summation of those of *A* and *B*. As we know another potential advantage of mixing solids is to increase the surface area of the solids in order to have faster reaction rates. Such a mixing scenario doesn't show too much advantage in shifting the capture temperature, but may enhance the kinetics of the capture process and eventually make the mixtures more efficient. Although there is no such report in literature, we think such an attempt is worthwhile and are working on several doped systems. Figure 5 shows a case of Na-/K- doped Li_2ZrO_3 capture CO_2 through the following reaction (b):[24]

$$LiMZrO_3 + CO_2 \leftrightarrow \frac{1}{2} Li_2CO_3 + \frac{1}{2} M_2CO_3 + ZrO_2 \qquad (M = Na, K) \qquad (b)$$

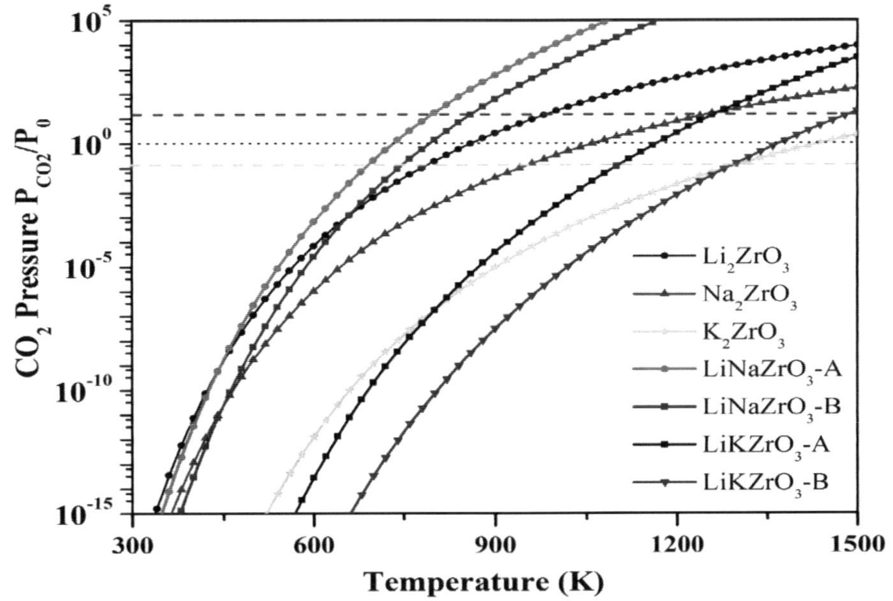

Fig. 5. The contour plotting of calculated chemical potentials versus CO_2 pressures and temperatures of the CO_2 capture reactions by $LiMZrO_3$ and M_2ZrO_3. Y-axis plotted in logarithm scale. Only the $\Delta\mu=0$ curve is shown explicitly. For each reaction, above its $\Delta\mu=0$ curve, their $\Delta\mu<0$, which means the solids absorb CO_2 and the reaction goes forward, whereas below the $\Delta\mu=0$ curve, their $\Delta\mu>0$, which means the CO_2 start to release and the reaction goes backward to regenerate the sorbents.

All of the reactions are thermodynamically favorable for CO_2 sorption over a quite wide range of temperatures and reduced CO_2 pressures. But a well preforming CO_2 solid sorbent should not only easily absorb CO_2 in the first half cycle but also easily release CO_2 from the product (M_2CO_3 and ZrO_2 for example) in the second half cycle. Table 1 shows the calculated thermodynamical properties and turnover temperatures of $LiMZrO_3$ capture CO_2 reaction (b).

From Fig.5 and Table 1, one can see that they are not good sorbents for capturing CO_2 for pre-combustion technology. However, they could be used for high-temperature post-combustion CO_2 capture with T_2=685K, 745 K for $LiNaZrO_3$ (phases _A and _B) and T_2=1095K, 1285 K for $LiKZrO_3$ (phases _A and _B) respectively. When compared to Li_2ZrO_3, the K substituted system $LiKZrO_3$ does not decrease the turnover temperature of the CO_2 capture reaction. In other words, although the $LiKZrO_3$ has a stronger driving force to capture CO_2, with improved CO_2 capture capability when compared to Li_2ZrO_3, its regeneration only can happen at higher temperature. Therefore, $LiKZrO_3$ is not a good CO_2 sorbent, since it requires more energy to regenerate. In contrast, the Na-substituted sorbent system $LiNaZrO_3$ does have a more compatible turn over temperature, making it a possible good candidate as a CO_2 sorbent. Such conclusions are in good agreement with experimental findings.[25-27]

In summary, from the calculated thermodynamic properties of $LiMZrO_3$ reacting with CO_2, we found that the CO_2 sorbent performance of $LiNaZrO_3$ is better than that of Li_2ZrO_3, but the performance of $LiKZrO_3$ is much worse since its regeneration temperature is higher. Therefore, the Na doped Li_2ZrO_3, $LiNaZrO_3$, is a better CO_2 sorbent for post-combustion capture technology. Our calculated thermodynamic results showed that the K doped Li_2ZrO_3, $LiKZrO_3$, didn't gain any improvements over Li_2ZrO_3 on its overall CO_2 capture performance.

Table 1. The weight percentage of CO_2 capture, the calculated energy change ΔE^{DFT}, the zero-point energy changes ΔE^{ZP} and the thermodynamic properties (ΔH, ΔG) of the CO_2 capture reactions. (unit: kJ/mol). The turnover temperatures (T_1 and T_2) of the reactions of CO_2 capture by solids under the conditions of pre-combustion (P_{CO2}=20 bar) and post-combustion (P_{CO2}=0.1 bar) are also listed.

reaction	absorbing CO_2 Wt%	ΔE^{DFT}	ΔE^{ZP}	ΔH (T=300K)	ΔG (T=300K)	Turnover T (K)	
						T_1	T_2
$Li_2ZrO_3+CO_2 \leftrightarrow Li_2CO_3+ZrO_2$ [a]	28.75	-146.648	11.311	-158.562	-103.845	1000	780
$LiNaZrO_3$_A+$CO_2 \leftrightarrow ½(Li_2CO_3+Na_2CO_3)+ZrO_2$	26.01	-152.936	7.069	-176.666	-110.892	805	685
$LiNaZrO_3$_B+$CO_2 \leftrightarrow ½(Li_2CO_3+Na_2CO_3)+ZrO_2$	26.01	-167.872	6.934	-191.526	-126.477	865	745
$LiKZrO_3$_A+$CO_2 \leftrightarrow ½(Li_2CO_3+K_2CO_3)+ZrO_2$	23.75	-264.115	6.006	-287.513	-225.611	1275	1095
$LiKZrO_3$_B+$CO_2 \leftrightarrow ½(Li_2CO_3+K_2CO_3)+ZrO_2$	23.75	-311.604	7.080	-332.612	-272.410	hT[b]	1285
$Na_2ZrO_3+CO_2 \leftrightarrow Na_2CO_3+ZrO_2$ [a]	23.76	-140.862	2.236	-158.327	-114.121	1275	925
$K_2ZrO_3+CO_2 \leftrightarrow K_2CO_3+ZrO_2$ [a]	20.24	-223.158	5.813	-238.490	-187.884	hT[b]	1285

[a] from Refs.[15, 19, 20].
[b] hT means the temperature is higher than our temperature range (1500K)

IV. Conclusions

The obtained results showed that by changing the mixing ratio of solid *A* and solid *B* to form mixed solid *C* it's possible to shift the turnover T_t of the newly formed solid *C* to fit the practical CO_2 capture technologies. When mixing SiO_2 or ZrO_2 into the strong Li_2O sorbent, one can obtain a series of lithium silicates (or zirconates) with T_t lower than that of pure Li_2O. By mixing oxides (Na_2O, K_2O, CaO) or their corresponding carbonates into MgO, the obtained mixtures exhibit different thermodynamic behaviors and their T_t are higher than that of pure MgO. Such results can be used to provide insights for designing new CO_2 sorbents. Therefore, although one single material taken in isolation might not be an optimal CO_2 sorbent to fit the particular needs to operate at specific temperature and pressure conditions, by mixing or doping two or more materials to form a new material, our results showed that it is possible to synthesize new CO_2 sorbent formulations which can fit the industrial needs. Our results also show that computational modeling can play a decisive role for identifying materials with optimal performance.

Acknowledgement

The author thanks Drs. D. C. Sorescu, D. Luebke, H. W. Pennline, B. Y. Li, H. Pfeiffer, J. K. Johnson, B. Zhang, K. Zhang, D. King, K. Parlinski for their kind help and collaborations.

References

1 B. Y. Li, Y. Duan, D. Luebke, and B. Morreale, Applied Energy **102**, 1439 (2013).
2 D. Aaron and C. Tsouris, Separation Science and Technology **40**, 321 (2005).
3 M. R. Allen, D. J. Frame, C. Huntingford, C. D. Jones, J. A. Lowe, M. Meinshausen, and N. Meinshausen, Nature **458**, 1163 (2009).
4 R. S. Haszeldine, Science **325**, 1647 (2009).
5 Y. Duan, D. Luebke, and H. W. Pennline, International Journal of Clean Coal and Energy **1**, 1 (2012).
6 Y. Duan and D. C. Sorescu, Phys Rev B **79**, 014301 (2009).
7 Y. Duan and D. C. Sorescu, J Chem Phys **133**, 074508 (2010).
8 Y. Duan, in *Proceedings of 7th~12th Ann. Conf. on Carbon Capture, Sequestration & Utilization*, Pittsburgh, 2008-2013).
9 Y. Duan and K. Parlinski, Phys Rev B **84**, 104113 (2011).
10 Y. Duan, B. Zhang, D. C. Sorescu, and J. K. Johnson, J. Solid State Chem. **184**, 304 (2011).
11 Y. Duan, D. R. Luebke, H. W. Pennline, B. Y. Li, M. J. Janik, and J. W. Halley, Journal of Physical Chemistry C **116**, 14461 (2012).
12 Y. Duan, B. Zhang, D. C. Sorescu, J. K. Johnson, E. H. Majzoub, and D. R. Luebke, Journal of Physics-Condensed Matter **24**, 325501 (2012).
13 B. Zhang, Y. Duan, and J. K. Johnson, J Chem Phys **136**, 064516 (2012).
14 Y. Duan, Phys. Rev. B **77**, 045332 (2008).
15 Y. Duan, J Renew Sustain Ener **3**, 013102 (2011).
16 Y. Duan, J Renew Sustain Ener **4**, 013109 (2012).
17 M. W. J. Chase, J. Phys. Chem. Ref. Data, **Monograph 9**, 1 (1998).
18 Y. Duan, Physical Chemistry Chemical Physics **15**, 9752 (2013).
19 Y. Duan, H. Pfeiffer, B. Y. Li, I. C. Romero-Ibarra, D. C. Sorescu, D. R. Luebke, and J. W. Halley, Physical Chemistry Chemical Physics **15**, 13538 (2013).
20 Y. Duan, J Renew Sustain Ener **4**, 013109 (2012).
21 K. L. Zhang, X. H. S. Li, Y. Duan, D. L. King, P. Singh, and L. Y. Li, International Journal of Greenhouse Gas Control **12**, 351 (2013).
22 K. L. Zhang, X. H. S. Li, W. Z. Li, A. Rohatgi, Y. Duan, P. Singh, L. L. Y., and D. L. King, Advanced Materials Interfaces (2014) accepted.
23 Y. Duan, K. L. Zhang, X. H. S. Li, D. L. King, B. Y. Li, L. F. Zhao, and Y. H. Xiao, Aerosol and Air Quality Research, DOI: 10.4209/aaqr.2013.05.0178 (2014).
24 Y. Duan, ScienceJet, submitted (2014).
25 M. Y. Veliz-Enriquez, G. Gonzalez, and H. Pfeiffer, J Solid State Chem **180**, 2485 (2007).
26 H. Pfeiffer, E. Lima, and P. Bosch, Chemistry of Materials **18**, 2642 (2006).
27 H. Pfeiffer, C. Vazquez, V. H. Lara, and P. Bosch, Chemistry of Materials **19**, 922 (2007).

351628 Utilization of Carbonic Anhydrase-Displaying Escherichia Coli in the Foam Bioreactor to Capture and Sequester Carbon Dioxide

Abstract:

In recent years, global warming as the result of greenhouse gas emissions, especially that of CO_2, has become of great concern. Many technologies have been proposed and employed to capture post-combustion CO_2. Conventional methods such as absorption and adsorption have the problems of regeneration of spent solvents and media requiring additional energy input and expensive operating costs. For these reasons, the carbonic anhydrase enzyme (CA), which quickly catalyzes the conversion of CO_2 into bicarbonate ion and a proton, has been studied for use in carbon capture and sequestration (CCS) technologies. This process would solve the limitations of current carbon capture technologies because it readily converts CO_2 into environmentally friendly products at mild conditions. Furthermore, upon CO_2 removal, the aqueous bicarbonate can be precipitated and separated as calcium carbonate (bio-mineral) as a value-added product. However, current CA-driven carbon capture remains expensive: the purified enzyme itself costs in the range of thousands of dollars per gram; the enzyme works best in slightly alkaline conditions, so some buffering capacity is necessary to combat the protons it generates; and CA must be replaced over time as it gradually loses its activity. The major goals of this project are to employ genetically modified *E. coli* to continuously produce and display CA on their cell membranes and to use these bacteria in the Foam Bioreactor to capture and sequester CO_2. The bacteria incubated in auto inducible media (AIM) have shown successful growth and enzyme expression as well as high activity and foaming capacity in the Foam Bioreactor. Because of the huge gas-to-liquid surface area in the Foam Bioreactor, the CO_2 in gas phase was quickly transferred to many CA-displaying bacteria, allowing the CA to convert CO_2 to bicarbonate ions and protons. Under *ex situ* pH control and an inlet stream of 0.5-2% CO_2 at 1 L/min, the increased cell concentrations enhanced the removal efficiency and elimination capacity of CO_2. A decrease in gas flow rate from 1.0 to 0.4 L/min at an inlet stream of 0.5-2% CO_2 also enhanced the removal efficiency of CO_2. Future work will include the effect of tertiary amine solutions (i.e., methyl diethanolamine) to enhance mass transfer and overall removal efficiency of CO_2, the *in situ* pH control in the foam for optimal CA activity inside the reactor, and the mathematical modeling of CA-driven carbon capture in the Foam Bioreactor.

352818 Purification of Waste Carbon Dioxide from Ethylene Glycol Plants and its' Commerical Application

Abstract:

Reduction of waste and convert them to the valuable product is the key challenge in today's competitive petrochemical business. Large amount of the carbon dioxide is treated as waste and vented to atmosphere from ethylene glycol plant. Carbon dioxide (CO_2) is one of the byproducts produced during the manufacturing process of ethylene glycol. Instead, it may be used as one of the raw material in urea & methanol process, soda ash, baking powder etc. Moreover, it can be liquefy up to food grade or can be utilized for dry Ice, cold drink, beverage preparation, even as preservative etc. These studies will emphasis on opportunity to reduce carbon dioxide emission to atmosphere and utilization in ecofriendly manner by conversion into competitive petrochemical products. This paper will also describes technology in detail, how to recover & purified the waste carbon di oxide to meet the specification of commercial & food grade as a raw material. The purification of waste CO_2 will provide two fold benefit. First, there will be reduction in the large CO_2 venting & subsequent reduction of pollution & global warming. Secondly, availability of raw materials from waste stream, which will reducing the production cost.

352865 Fuel Based Approach for CO2 Emission Factor Estimation

Abstract:

Global warming is an issue facing the world nowadays and CO_2 emissions are contributing directly to this issue. There are many ways to quantify and predict CO_2 emissions from different sources. Combustion process is considered to be the major source of CO_2 emissions. There are many published CO_2 emission factors in order to find the released emissions. In this study, CO_2 emission factors from fuel specifications for oil and petrochemical industry is used and compared with the published data. It was found that CO_2 emissions vary with quality of fuel and operating condition. For example, the average CO_2 emission factor for gas turbine was calculated as 114 lb/MMbtu. This emission factor has a good match with published emission factor of EPA and CAPP data.

350799 Adsorption of CO2 and H2S from Natural Gas by Activated Fly Ash

Abstract:

Waste oil fly ash (OFA), a by- product of fossil fuel combustion from power generating plants, was successfully convert to activated carbon using both chemical and physical treatment. A combination of three acids (Sulfuric, nitric and phosphoric acids) was mixed at different volume ratios and concentrations with the ash, followed by physical treatment at 990 oC while CO_2 is introduced at flow rate of 1 L/min. The produced AC was characterized according to its surface area and morphology. BET surface area has increased from less than 1 m^2/g for untreated ash to 375 m^2/g after both chemical and physical activation. Then it is utilized in the removal of CO_2 and H_2S from natural gas stream. The maximum adsorption capacity of the treated ash was found to be 240 mg/g and 175 mg/g, respectively. The adsorption rate increased with the increase in flow rate and concentration. The results indicated the potential of using waste OFA in acidic gases removal from downstream.

353458 Common Issues that Can Limit Steam Trap Performance

Abstract:

The proper selection of steam trap type/technology for a given application is just one of the keys to achieve operational success. In this session we will discuss a variety of factors such as piping condition, piping arrangement, and steam trap installation/orientation/elevation that can also limit steam trap performance. Common application problems like steam locking and group trapping will also be discussed.

353457 Overview of Steam Trap Types and Their Operation

Abstract:

Steam traps are the quietly functioning devices underneath the pipe racks of our manufacturing facilities, giving the occasional breath of condensate/steam to the atmosphere. However they are not all created equally or function the same way. This presentation will give an overview of various types of steam traps, how they operate, and review steam trap selection for common applications.

A Practical Approach to Feedwater Quality Requirements in Boiler Systems

Paper No. 352255
Deborah Bloom, Principal Consultant, Nalco, an Ecolab company, Naperville, IL
David Puchan, Senior Industry Technical Consultant, Nalco Champion, an Ecolab company, Sugar Land, TX

ABSTRACT

The steam boiler system is a critical component of most manufacturing processes. Steam production reliability is often a key component in assuring product quality and overall production efficiency. Mechanical, operational, and chemical requirements of the steam will usually determine what boiler design and feedwater or makeup water quality is needed. If you define the steam requirements, you typically define the rest of the system.

This paper will discuss typical purity and quality requirements for some common steam users and show how that translates to boiler design, feedwater and makeup purity requirements, and condensate return. We will also discuss how typical treatment program choice, chemical control limits, and boiler blowdown rates are affected by steam purity requirements.

INTRODUCTION

Mechanical, operational, and chemical requirements for the steam will typically determine what boiler design and feedwater/makeup water quality are needed by the boiler system. Key questions to ask include, but are not limited to, the following:

- Are there steam purity or quality specifications that need to be met?
- How is the steam used? For example, does the system include turbines, superheaters, sterilizers, humidifiers, direct injection processes, or indirect heat transfer?
- What are the required temperature and pressure characteristics of the steam?
- How much steam will be needed and is the need continuous or intermittent?
- Are there government regulations or other regulations that need to be met?

Answers to these key questions will help define steam requirements. If you define the steam requirements, you will typically define the rest of the system - with one exception. Waste heat boilers may have more stringent feedwater quality requirements than are warranted by the pressure/temperature of the steam they produce.

Treatment program choice, chemical control limits, and boiler blowdown rates are also affected by steam purity requirements.

STEAM PURITY VS. STEAM QUALITY

Steam purity is defined as the amount of non-water components contained in the steam. These components can be dissolved in the steam, in water droplets entrained in the steam, or carried as discreet solid particles with the steam. Steam purity in industrial systems is normally measured in terms of parts per billion (ppb) sodium ion or silica. Gaseous components such as ammonia, amine, and carbon dioxide are not normally included in this expression, but there can also be limitations on these in the steam.

Steam quality is defined as the amount of moisture in the steam, with 100% quality equal to 100% dry steam, i.e. no moisture. Moisture from the boiler will contain boiler dissolved solids which can be harmful to the steam user. In addition, the moisture itself can be erosive to turbine inlet nozzles and blades, pipe elbows, and valves. For industrial boilers that typically operate at less than 2000 psig drum pressure, the primary route for inorganic contaminants, other than silica, getting into the steam is various forms of mechanical carryover - droplets of boiler water entrained in the saturated steam, including foaming. Therefore, the boiler water total dissolved solids (TDS) are directly proportional to the saturated steam TDS. Only silica has any appreciable vaporous solubility in saturated steam at these operating pressures, and its specific concentration in the boiler water must be appropriately controlled to limit its concentration in the saturated steam when drum operating pressures are above 600 psig. Typical desired steam silica concentrations are less than 10-20 ppb (as SiO_2).

Superheated steam is generally required for turbines and turbine drives to prevent damage by moisture and steam entrained contaminants. Superheated steam has been heated to a temperature significantly greater than the saturation temperature of the operating pressure and therefore by definition has no moisture. Appreciable dissolved solids entering the superheater in the saturated steam can cause a buildup of deposits in the superheater elements, often leading to overheating and tube failures. As a result, a low total dissolved solids content is desirable for steam going to an industrial superheater. Superheaters vary considerably in design but can nominally tolerate 100-200 ppb of total dissolved solids in the steam. This is equivalent to about 33-67 ppb of sodium (as Na) in the steam (1).

Superheaters, turbines and turbine drives normally have the highest steam purity restrictions among common steam users. Superheaters can tolerate higher levels of impurities in the steam prior to experiencing deposition problems. Acceptable steam purity for an industrial, condensing steam turbine is typically less than 10 ppb sodium (as Na) in the steam and less than 20 ppb sodium (as Na) in the steam for an industrial backpressure turbine. Some turbine designs are more sensitive to deposition than others and new turbine installations may have even more restrictive steam purity guarantee limits. Table 1 shows the chemistry limits from several manufacturers for condensing steam turbines operating under normal conditions. High purity feedwater/ makeup is typically required to meet these steam restrictions, as well as the purity requirements for use as attemperation water prior to a turbine. High purity feedwater/makeup can usually be achieved by polished or mixed bed demineralization, polished reverse osmosis, or a high percentage of polished condensate (greater than 90%).

TEMPERATURE AND PRESSURE REQUIREMENTS AFFECT BOILER DESIGN

Temperature and pressure requirements for steam users affect boiler design and feedwater quality as well. Higher pressure steam users (greater than 600 psig) require higher pressure boilers, while turbines and turbine drives usually require superheated steam and must therefore be provided by a boiler with a superheater. Integral superheaters are typically limited to water tube boilers, heat recovery steam generators, and waste heat boilers.

Knowledge of the boiler design – firetube, water tube, heat recovery steam generator, or waste heat boiler, as well as direct fired vs. heat recovery – will help in understanding potential problems such as scaling and corrosion. By knowing the design of the boiler, one can anticipate where these problem areas are located and what may have caused them, such as:

- Water circulation issues
- Boiler blowdown location and limitations
- High heat flux areas

Table 1. Comparison of Manufacturers' Chemistry Limits for Condensing Steam Turbines Operating under Normal Conditions (2)

Steam (HP, IP, LP)	Alstom	GE[a]		Mitsubishi	Fuji	Siemens
		Reheat	No reheat			
Specific conductivity at 25°C, μS/cm	3-11	NS	NS	NS	NS	NS
Cation conductivity at 25°C, μS/cm	< 0.2	≤ 0.15	≤ 0.25	< 0.3	<0.3	< 0.2
pH at 25°C	9.0-9.6 [b]	NS	NS	NS	9.0-9.3	NS
Silica as SiO_2, ppb (μg/kg)	<20	≤ 10	≤ 20	< 10	<10	<10
Total iron as Fe, ppb (μg/kg)	<20	NS	NS	< 20	<20	<20
Total copper as Cu, ppb (μg/kg)	<3	NS	NS	< 2	<3	<2
Sodium as Na, ppb (μg/kg)	<10	≤ 3	≤ 6	< 5	<10 [c]	<5
Chloride as Cl, ppb (μg/kg)	NS	≤ 3	≤ 6	< 5	NS	NS
Sulfate as SO_4, ppb (μg/kg)	NS	≤ 3	≤ 6	NS	NS	NS
Oxygen as O_2, ppb (μg/kg)	NS	NS	NS	< 10	<10	NS
TOC as C, ppb (μg/kg)	NS	≤ 100	≤100	NS		NS

[a] *Using phosphate treatment (greater than 2.6 Na:PO_4 molar ratio) or AVT*
[b] *pH 9.0 – 9.3 with copper-alloy condensers*
[c] *Na + K*
NS is not specified

Temperature and pressure requirements also affect the boiler chemistry, which in turn affect feedwater and makeup quality.

BOILER SYSTEM CHEMISTRY AND CONTROL LIMITS

There are many water-related challenges encountered when operating a boiler. Two of the most common problems are scale or deposition and corrosion.

Scale is the result of dissolved minerals, such as calcium and magnesium, exceeding their solubilities and forming deposits (3). Boiler heat transfer efficiency can be negatively impacted by scale. In addition porous deposits formed by metal oxides (typically iron and copper) can also create serious under-deposit corrosion problems. Deposition, when not addressed in a timely fashion or when improperly dealt with, can result in ruptured boiler tubes from overheating or other boiler damage. This can cause boiler shutdown and lost or greatly reduced production.

Most boiler systems follow some sort of published guidelines to establish a suitable control strategy for their feedwater and boiler water chemistry. Equipment manufacturers, water treatment suppliers, American Boiler Manufacturers' Association (ABMA), and American Society of Mechanical Engineers (ASME) are among the more common sources for such guidelines. Table 2 shows an example of the recommended control limits provided by ASME (4). Note that chemistry limits generally become more restrictive for both feedwater and boiler water contaminants as system pressure increases. The table assumes that only demineralized makeup is used above 1000 psig. Waste heat boilers of unusual design and heat recovery steam generators (HRSGs) will frequently require more stringent control limits.

Table 2. ASME Suggested Water Chemistry Limits for Industrial Water Tube, High Duty, Primary Fuel Fired, Drum Type Boilers
Makeup water percentage: Up to 100% of feedwater
Conditions: Includes superheater, turbine drives, or process restriction on steam purity
Saturated steam purity target: See tabulated values below

Drum Operating Pressure[a,k]	**psig 0-300 (MPa) (0-2.07)**	**301-450 (2.08-3.10)**	**451-600 (3.11-4.14)**	**601-750 (4.15-5.171)**	**751-900 (5.18-6.211)**	**901-1000 (6.22-6.89)**	**1001-1500 (6.90-10.34)**	**1501-2000 (10.35-13.79)**
Feedwater[g]								
Dissolved oxygen ppm (mg/L) O_2 measured before chemical oxygen scavenger addition[h]	< 0.007	< 0.007	<0.007	<0.007	<0.007	<0.007	<0.007	<0.007
Total iron ppm (mg/L) Fe	≤0.1	≤0.05	≤0.03	≤0.025	≤0.02	≤0.02	≤0.01	≤0.01
Total copper ppm (mg/L) Cu	≤0.05	≤0.025	≤0.02	≤0.02	≤0.015	≤0.01	≤0.01	≤0.01
Total hardness ppm (mg/L)*	≤0.3	≤0.3	≤0.2	≤0.2	≤0.1	≤0.05	ND	ND
pH @ 25°C	8.3-10.0	8.3-10.0	8.3-10.0	8.3-10.0	8.3-10.0	8.8-9.6	8.8-9.6	8.8-9.6
Chemicals for preboiler system protection	NS	NS	NS	NS	NS	VAM	VAM	VAM
Nonvolatile TOC ppm (mg/L) C[f]	<1	<1	<0.5	<0.5	<0.5	<0.2	<0.2	<0.2
Oily matter ppm (mg/L)	<1	<1	<0.5	<0.5	<0.5	<0.2	<0.2	<0.2
Boiler Water								
Silica ppm (mg/L) SiO_2	≤150	≤90	≤40	≤30	≤20	≤8	≤2	≤1
Total alkalinity ppm (mg/L)*	<700[c]	<600[c]	<500[c]	<200[c]	<150[c]	<100[c]	NS[d]	NS[d]
Free OH alkalinity ppm (mg/L)*[b]	NS	NS	NS	NS	NS	NS	ND[d]	ND[d]
Specific conductance[L] µmhos/cm (µS/cm) 25°C without neutralization	5400-1100[e]	4600-900[e]	3800-800[e]	1500-300[e]	1200-250[e]	1000-200[e]	≤150	≤80
Total Dissolved Solids in Steam[i]								
TDS (maximum) ppm (mg/L)	1.0-0.2	1.0-0.2	1.0-0.2	0.5-0.1	0.5-0.1	0.5-0.1	0.1	0.1

*as $CaCO_3$
NS = Not Specified
ND = Not Detectable
VAM = Use only volatile alkaline materials up stream of attemperation water source

Maximum boiler water silica concentrations at pressures above 600 psig are selected so that the volatile carryover will not exceed 20 ppb silica (as SiO_2) in the steam according to well established silica volatility data (5). At lower operating pressures ranges, the boiler water silica concentrations are selected to avoid internal silicate deposits.

Recommended boiler water alkalinity and conductivity values also decrease as boiler pressure increases. Recommended boiler water total alkalinity values were set based on best correlation with boiler water pH, corrosion inhibition, and carryover tendency. Hydroxide alkalinity values are usually set for a specific system based on boiler design, pressure and treatment program.

Boiler water conductivity is the typical trigger for a blowdown control event, although silica and alkalinity should be considered reasons to blow down as well.

Feedwater total hardness, iron, and copper are specified to minimize potential boiler deposits. The decrease in recommended concentration relative to increased boiler pressure reflects the corresponding potential impact of the resultant deposits on heat transfer at higher pressures. Feedwater quality, and therefore makeup quality, dictates what the boiler blowdown rate must be to maintain the required boiler water chemistry limits. Higher blowdown rates result in increased water and energy loses, but are necessary to prevent deposition and corrosion. Some of the more common ways to prevent deposition from occurring in boilers include ion exchange (softening or demineralization) and reverse osmosis. Residual dissolved minerals are prevented from forming deposits by the addition of chemical treatment.

Table 2 also provides a connection between boiler chemistry and expected total dissolved solids in the steam. As boiler pressure increases, the specific conductivity required in the boiler water to achieve acceptable steam purity decreases. What's acceptable depends on how the steam will be used, but is typically 0.1 to 0.2 ppm TDS for superheater steam. Even this, however, is usually considered too high for most modern industrial turbines. ASME is the process of updating Table 2.

WASTE HEAT BOILERS

The design for a waste heat boiler generally depends on the process operating parameters. It is generally a one-of-a-kind unit, designed to accommodate and facilitate a plant's process flows. In flue gas waste heat recovery, where the process streams are low pressure, dirty, and require supplemental firing, watertube designs predominate. In process cooling applications, where the processes are corrosive and at relatively high pressure, firetube or shell-and-tube waste heat boiler designs are common.

Waste heat boiler design can cause limitations in flow, higher heat flux areas, limited or ineffective blowdown from areas where solids collect, poor circulation, and inadequate moisture separation from the steam. These problems can then lead to concentration mechanism corrosion, such as caustic gouging and phosphate wastage, creep rupture from overheating due to scale/deposits, and poor steam purity. System chemistry limits may need to be even lower than those shown in Table 2. For example, Table 3 shows recommendations for an Ethylene plant TLX/TLE feedwater chemistry; boiler water chemistry is specific to the treatment program that's in use.

Table 3. Boiler Feedwater Recommendations for Ethylene Plant TLX/TLE Operations (6)

Contaminant	Recommended Control Limit
Cation Conductivity	< 0.5 µmhos
Silica (as SiO_2)	< 20 ppb
Sodium (as Na)	< 10 ppb
Chlorides (as Cl)	< 6 ppb
Sulfates (as SO_4)	< 6 ppb
Calcium (as Ca)	Non-detectable
Magnesium (as Mg)	Non-detectable
Iron (as Fe)	< 10 ppb
Copper (as Cu)	< 2 ppb
TOC (as C)	< 100 ppb
Oxygen (as O_2)	< 2 ppb
pH	9.0 – 9.3 systems with yellow metals
pH	9.2 – 9.6 systems without yellow metals
Oil & Grease	<0.2 ppm

A FEW MORE WORDS ABOUT WATER QUALITY

By now it should be apparent that steam purity and temperature/pressure requirements determine boiler design, pressure, and chemistry. This in turn affects feedwater quality requirements, which affect makeup and condensate quality requirements. So, as mentioned initially, if you define the steam requirements, you will typically define the rest of the system.

Improving makeup water quality reduces the amount of feedwater contaminants entering the boiler system. There are a variety of pretreatment technologies available for industrial plants that can significantly reduce the hardness present in makeup:

- Hot/cold lime softening
- Sodium zeolite softeners
- Reverse Osmosis
- Demineralization

Table 4. Reverse Osmosis vs. Competing Technologies

Technology	Driver	Competing technology preferred if:	RO preferred if:
Hot-lime softener, filtration, sodium softening	Water quality	Need to remove hardness, alkalinity and silica only	Need to remove NaCl, TDS, alkalinity and silica to low concentrations
Sodium softeners	Water quality	Need to remove hardness only	Need to remove NaCl, TDS, alkalinity and silica to low concentrations
Dealkalizers	Water quality	Need to remove alkalinity only	Need to remove NaCl, TDS, alkalinity and silica to low concentrations
Demineralization	Cost	Cost of acid and caustic is low	Waste treatment cost high - To avoid handling of acid & caustic

Each provides a different effluent quality. Table 4 provides a comparison of these methods relative to reverse osmosis, including the drivers for choosing one technology over another.

CHEMICAL TREATMENT

The goal of chemical treatment is to optimize system chemistry and control in order to save water and energy. Left untreated, water in a boiler system will scale and corrode - even with the perfect pretreatment and chemistry control limits. Effective chemical treatment assures that the boiler system operates at maximum efficiency and reliability.

Scale and Deposition

Scale and deposit control is focused on preventing the precipitation of residual dissolved minerals remaining in the water after pretreatment, such as calcium and magnesium, from forming deposits in the boiler. Typical chemical treatment consists of using either a precipitating type program, such as phosphate, or a solubilizing type program, such as all-polymer. Both program types are fed relative to the amount of hardness still present in the feedwater. Phosphate programs can be used with very high feedwater hardness, but since the amount of precipitate formed in the boiler is in direct proportion to the hardness concentration, they also typically require higher blowdown rates to purge the sludge and minimize boiler scale formation or periodic mechanical or chemical cleaning to remove accumulated deposits.

There are many forms of all-polymer treatment in the market today. All can be effective at scale prevention when fed at the proper dosage relative to hardness. Newer versions of all-polymer programs tend to tolerate hardness upsets better than the original acrylate all-polymer programs. Boiler blowdown rates will be lower on all-polymer than with phosphate treatment, saving energy and water; and since all-polymer is a solubilizing type treatment, internal boiler surfaces will be cleaner which may result in additional fuel savings from improved heat transfer.

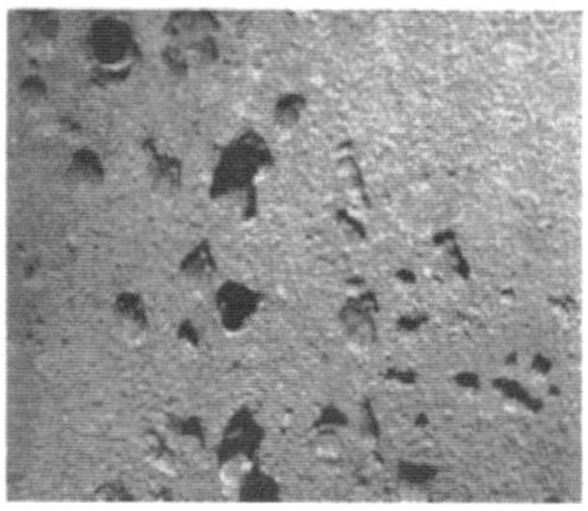

Figure 1. Oxygen attack, characterized by its classic pitting appearance

Corrosion

Corrosion is a serious concern in the pre-boiler and boiler system for two reasons. First, it shortens the asset life of the boiler and pre-boiler system. Second, the corrosion by-products (iron and copper oxides) resulting from unchecked corrosion can deposit in high heat transfer areas of the boiler as deposition. Oxygen corrosion is a specific and particularly devastating type of corrosion, Figure 1. It is a common problem that adversely affects the life of the key steam system components. Pitting attack from oxygen corrosion can cause premature metal failure.

Iron oxide deposits, just as the earlier described calcium and magnesium scales, reduce heat transfer efficiency and can result in under deposit corrosion. In uncontrolled scenarios, blown tubes (or other extensive boiler damage), boiler shutdowns and production losses can result.

Historically, the best practice for preventing oxygen corrosion in industrial boiler and pre-boiler systems has been mechanical deaeration for primary oxygen removal, followed by chemical oxygen scavengers to remove remaining trace quantities. Sulfite is typically the scavenger of choice in most low pressure systems, especially at pressures less than 250 psig. Sulfite adds solids to the system and is traditionally dosed to a prescribed residual in the boiler water. Non-sulfite scavengers also exist. In addition to scavenging oxygen, the non-sulfite programs will typically passivate metal surfaces further reducing corrosion and iron pickup, and do not contribute to boiler solids since most are volatile. As a result, non-sulfite scavengers may also reduce boiler blowdown requirements.

MONITORING AND CONTROL

For decades, the traditional practice employed to control many boiler water scale and oxygen control chemistries has been by sampling the blowdown. One primary reason for this was that the boiler concentrated the treatment chemicals sufficiently to enable them to be measured readily. Although modernization and improvement in test methods have enabled feedwater testing (e.g., cycling up is not required to run many of the tests), many operations hold onto the old practices of the 1930s through 1970s and test only in the blowdown. The major drawbacks resulting from this approach are related to the lag time between changes in feedwater conditions and those changes showing up in the blowdown.

In a blowdown based control scenario, the operator would sample the blowdown, perform an array of chemical tests, and adjust the feed rate of the scale inhibitor or the oxygen scavenger. This constant "add chemical - sample – test – adjust" control loop was the standard. This meant that if a test was missed, an opportunity for adjustment was also missed.

In addition to the challenge of the above testing loop, the physical capacity of the boiler often meant that a negative change in feedwater conditions might go completely undetected or not have a notable impact on the boiler blowdown composition for hours or even days. Feedwater conditions that were potentially adverse to system integrity (e.g., an oxygen excursion, a reduction in feedwater pH, or a badly contaminated condensate return stream) would only be seen after they had become history, if they were detected at all, because only the boiler water was being tested. What was missing was a proactive approach to protect the boiler as opposed to a reactive approach that adjusts after damage has occurred.

Ideally, one would want to measure and control boiler chemistry in the feedwater, because that would enable one to make adjustments to changing conditions in real-time, rather than chasing events or missing them entirely and suffering the outcome. The Nalco boiler feedwater treatment automation system allows us to do this. It is a combination of two different technologies. First is the patented solid state fluorometer, which is used to control boiler internal treatment (7, 8, 9), Figures 2 and 3. Nalco has used inert fluorescent tracers for system diagnostics and dosage control of boiler system treatment chemicals for many years. The fluorometer allows us to measure these materials at the part per billion level of detection.

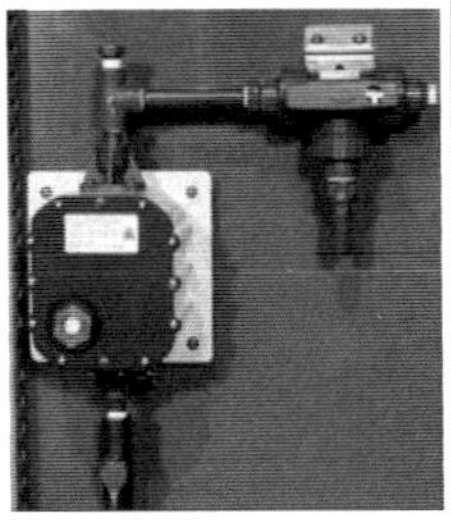

Figure 2. LED fluorometer provides continuous monitoring

Figure 3. Pen fluorometer used with grabs samples

This unique tracer technology makes it possible to determine the mass balance of a boiler system with greater accuracy, Figure 4. It also makes it possible to accurately and automatically control and monitor system chemistry and operating conditions, respond to boiler system changes, and maintain optimum treatment levels of highly effective scale and deposit inhibitors to keep boilers clean. As an example, this technology has been used in conjunction with soluble iron (Fe^{+2}) monitoring in a boiler system to quantify flow-assisted corrosion rates.

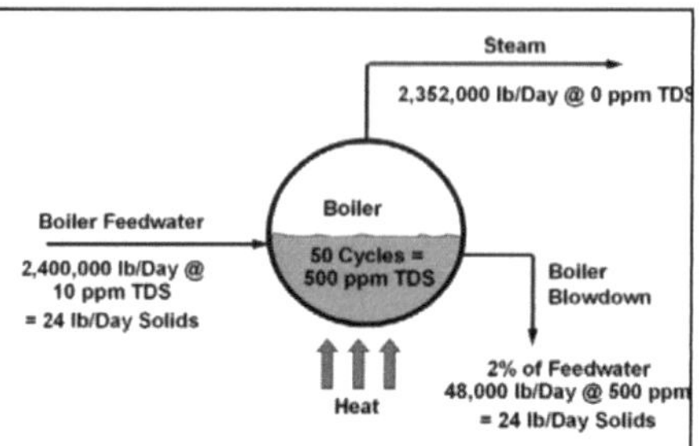

Figure 4. Boiler cycles of concentration

The second automation technology is the patented Nalco Corrosion Stress Monitor™, Figure 5. The Nalco Corrosion Stress Monitor instrument works by measuring the oxidation reduction potential at the temperature and pressure of the boiler feedwater. The At Temperature Oxidation Reduction Potential, or At Temperature ORP, detects changes in the corrosivity of the feedwater and can be used to control the oxygen scavenger feed. The AT ORP has a design rating up to 500°F and 2,800 psig. A uniquely designed controller controls both the fluorometer and AT ORP. Monitoring, On/Off Control, and PID (Proportional-Integral-Derivative) Control are all possible with this controller. These technologies are capable of communicating via web or DCS for monitoring and alarming existing conditions.

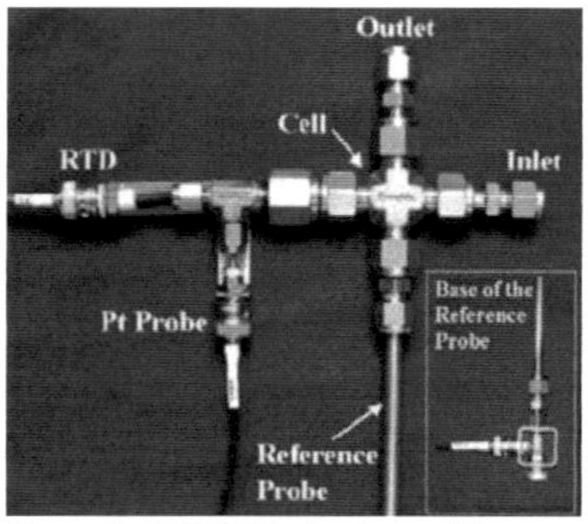

Figure 5. AT ORP Probe Assembly

OPPORTUNITIES FOR ENERGY SAVINGS

Controlling boiler deposition can have real energy savings benefits. There are many different sources of boiler deposition – two of the most common are feedwater hardness and dissolved iron entering the boiler feedwater.

The new control technologies can have profound impact on minimizing the potential for deposition, as well as mitigating its impact. Since the deposit inhibitor is measured in the boiler feedwater and adjusted in real time, regardless of changes in steam load, underdosing and overdosing can be almost completely eliminated. Boiler feedwater corrosivity is likewise monitored and controlled in real-time, so feedwater corrosion excursions, and the resultant deposit-forming dissolved iron can also be greatly reduced.

Other opportunities for energy savings spring from these technologies. It is important to remember that sulfite, one of the most common oxygen scavengers, contributes dissolved solids to the boiler feedwater. These solids increase boiler water conductivity. In a scenario where excessive levels of sulfite oxygen scavenger are being fed, improved control of oxygen scavenger feed can result in reduced boiler feedwater dissolved solids, higher cycles of concentration in the boiler, and reduced blowdown, all with associated water and fuel savings.

This technology also has the capability of monitoring and controlling feedwater pH, so energy and water savings from reduced feedwater dissolved solids, higher boiler cycles, and reduced blowdown can also be obtained by facilities that use caustic (sodium hydroxide) for controlling feedwater and boiler water alkalinity.

CASE HISTORY - A WESTERN U.S. REFINERY OPTIMIZES PERFORMANCE with 3D TRASAR™ BOILER AUTOMATION

Situation

A western U.S. refinery was experiencing deposition in their 600 psig boiler superheater tubes. This deposition forced operations to conduct frequent water washes, limit steam production, and operate the boiler at only 5 cycles resulting in an increase in energy, water, and maintenance costs. The superheater tube deposition led to very high skin temperatures and ultimately, tube failure.

Refinery management asked Nalco to help them understand their boiler system better. Nalco performed an Engineering Approach Analysis to determine system gaps and opportunities for improvement.

A Nalco Engineering Approach Analysis

Boiler Feed Water Quality - The 600 psig boiler system receives a combination of deaerated condensate return and reverse osmosis water that passed through a Graver Hot Process Softener used only for deaeration. Figure 6 shows a schematic of the boiler feedwater system.

Figure 6. Schematic of Boiler Feedwater System

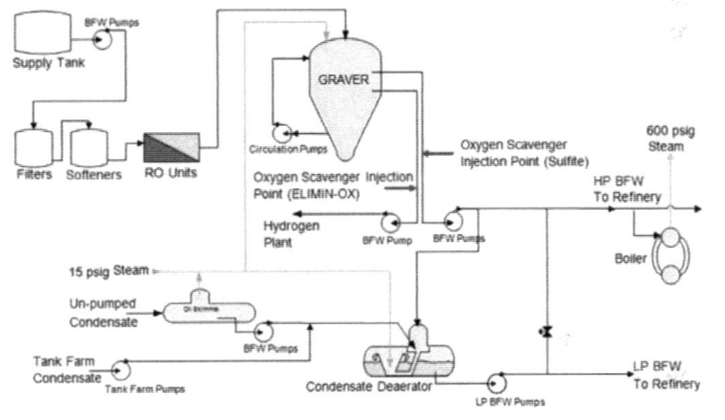

The feedwater from the Graver unit splits into two streams -one stream supplies feedwater to the refinery while the other stream supplies feedwater to a hydrogen plant. At the time of the initial audit, two different oxygen scavengers were fed into each of the boiler feedwater lines up-stream of the boiler feed water pumps. Sulfite was added to the refinery's feedwater line and erythorbate was injected into the feedwater line supplying the hydrogen plant. Sulfite cannot be used in hydrogen plants due to the potential for the product to poison the process unit's catalyst.

Figure 7. Manual Oxygen Testing of the Feedwater

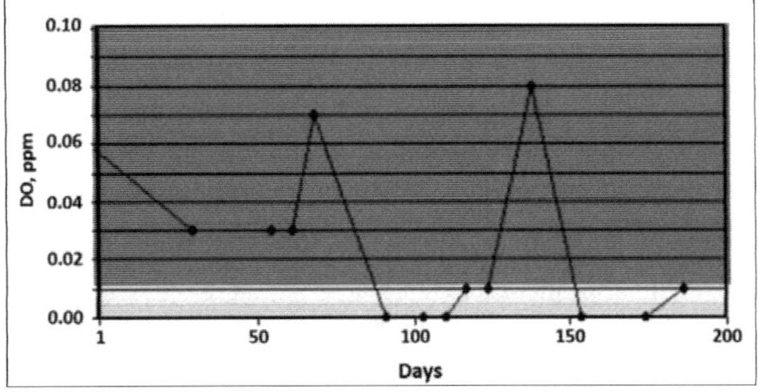

We suspected that the Graver was not mechanically removing the appropriate amount of dissolved oxygen. Manual testing confirmed this suspicion, Figure 7. Dissolved oxygen concentration varied, but at times was as high as 0.08 ppm or 80 ppb, which is well above industry best practices.

Plant operations fed tremendous amounts of sulfite to try to scavenge the high oxygen concentrations. Since the sulfite product is acidic, additional caustic was also required to help maintain feedwater pH. As more and more sulfite and caustic were added to the feedwater, the boiler water began to carryover. The solids in the boiler water deposited on the superheater tubes, increasing the outside metal tube temperature, and resulting in overheating and tube failure, Figure 8.

Figure 8. Superheater Tube Temperature

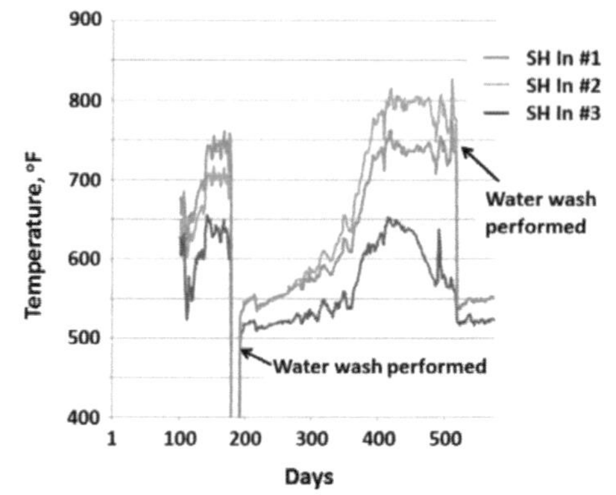

Recommendations

After careful review of the customer's system, chemical treatment program, and daily operations via the Engineering Approach Analysis, Nalco made several recommendations:

- Inspect and evaluate the boiler's internal steam separation equipment.
- Consider replacing the Graver with a traditional deaerator capable of mechanically reducing dissolved oxygen levels to less than 10 ppb.
- Evaluate whether a non-sulfite oxygen scavenger could be used.
  - Switching to a non-sulfite scavenger would reduce solids contribution to the feedwater and the potential for boiler carryover, thereby reducing the potential for deposition in the superheater tubes.
  - Switching scavengers would also potentially allow higher cycles of concentration in the boiler, reducing the energy and water usage.
  - Using an oxygen scavenger compatible with the hydrogen plant's requirements would allow the plant to use a single oxygen scavenger, which could then be injected into the Graver unit allowing longer reaction time.
- Utilize Nalco 3D TRASAR Boiler Automation to continuously monitor and control the chemical treatment and other critical steam system parameters based on system stress and demand.

Results

The decision was made to implement the Nalco recommended plan using a gradual approach. Soon after several of these actions were taken, the refinery experienced dramatic improvements.

Boiler Feed Water Quality - The 3D TRASAR Boiler technology confirmed that very high concentrations of dissolved oxygen were present in the feedwater, Figure 9.

Figure 9. Feedwater Dissolved Oxygen vs. Corrosion Potential

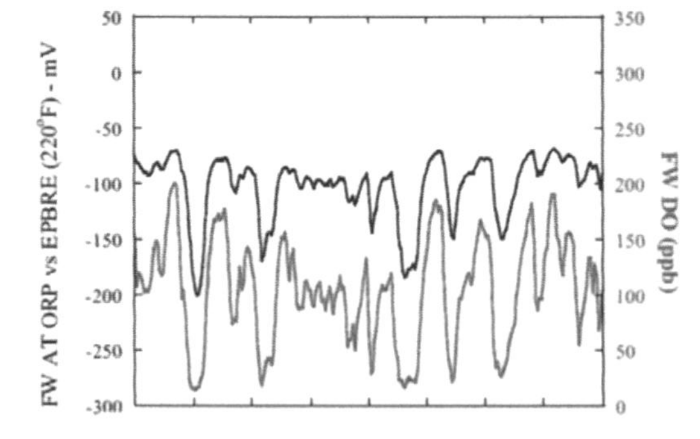

The refinery agreed to replace the Graver unit with a traditional deaerator as a long term solution, however in meantime, they needed a plan to reduce corrosion potential.

Erythorbate and carbohydrazide scavengers were tested as a replacement the existing sulfite application. Both met Hydrogen plant requirements, so the scavenger injection point was moved inside the Graver unit to improve mixing and residence time. Both scavengers performed well at removing dissolved oxygen, but carbohydrazide was the preferred chemistry because it had less effect on feedwater pH at high dosages and does not contribute dissolved solids like sulfite.

Once the best oxygen scavenger and injection point were in place, the next step was to put scavenger feed into automatic control mode via the Nalco Corrosion Stress Monitor. Figures 10 and 11 show time periods from both monitoring and control mode. Although oxygen concentrations and corrosion potential still spikes occasionally due to a sudden increase in dissolved oxygen, the system is quickly brought back into control range.

The AT ORP set point was -350 mv, reflecting system conditions. Once in control mode, ***control of corrosion potential improved from 50% on target to 91%.***

The real proof that 3D TRASAR Automation was improving boiler system reliability was the in superheater tube temperature data, Figure 12. Remember that boiler water was carrying over into the steam and depositing in the superheater tubes. The refinery had to shut down the boiler to water wash the superheater tubes to prevent failure from overheating. However with Nalco's action plan in place, the superheater tube temperatures have remained on target.

Summary

The refinery's utility area did not have a traditional deaerator, but rather a poorly operating Graver unit that resulted in excessively high feedwater dissolved oxygen concentrations. The 3D TRASAR Boiler Automation with the Nalco Corrosion Stress Monitor detected and responded immediately to changes in feedwater oxygen and corrosion stress conditions caused by the poorly operating deaerator. Manual feed of an oxygen scavenger provides poor feedwater corrosion stress control. The use of 3D TRASAR Technology to automatically control the feed of scavenger significantly increased percent on target control and reduced feedwater corrosion. Feeding carbohydrazide, a non-sulfite oxygen scavenger, reduced the overall solids loading on the boiler, eliminated boiler water carryover, and enabled the plant to maintain safe superheater tube temperatures. Because of the reduced solids loading, the refinery was also able to safely

Figure 10. Feedwater Corrosion Control using the Nalco Corrosion Stress Monitor

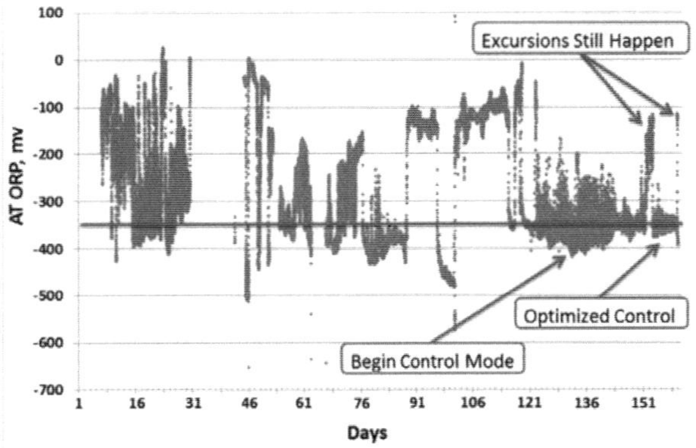

Figure 11. Feedwater Corrosion Control Showing Dissolved Oxygen

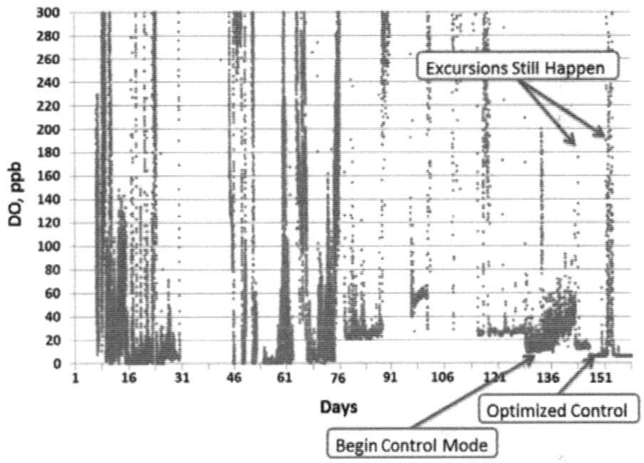

Figure 12. Nalco's Engineered Approach Analysis and 3D TRASAR Automation resulted in stable superheater tube temperatures and improved system reliability

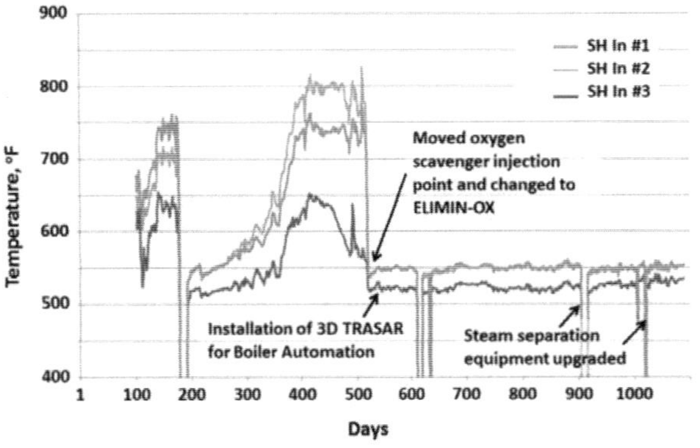

increase the boiler cycles, yielding $211,000 in annual savings.

CONCLUSIONS

Mechanical, operational, and chemical requirements of the steam will usually determine what boiler design and feedwater or makeup water quality is needed. If you define the steam requirements, you typically define the rest of the system. Typical treatment program choice, chemical control limits, and boiler blowdown rates are also affected by steam purity requirements. New control technologies can have profound impact on minimizing the potential for scale, deposition, and corrosion in the boiler system, as well as mitigating their impact. Water and fuel savings from reduced feedwater dissolved solids, higher cycles of concentration in the boiler, and reduced blowdown can be achieved.

REFERENCES:

1. "Industrial Steam Purity: Requirements, Proper Sampling and Practical Considerations", A Banweg, Paper IWC-08-28, Proceedings of the International Water Conference (2008).
2. ASME, "Consensus on Operating Practices for Control of Water and Steam Chemistry in Combined Cycle and Cogeneration Power Plants" (Jan 2012).
3. Nalco Water Handbook, 3rd Edition, 2009, McGraw Hill Publishing
4. ASME, "Consensus on Operating Practices for the Control of Feedwater and Boiler Water Chemistry in Modern Industrial Boilers" CRTD-Vol. 34 (1994).
5. "Selective Silica Carryover in Steam," E. Coulter, E. Pirsh, and E. Wagner, Jr., Transactions of ASME, 78:869-873 (1956).
6. "Boiler Technology to Monitor and Control Corrosion In Ethylene Plant Steam Generation Process", D. Puchan and M. Gerty, AIChE Paper Number 49 G, AIChE Spring National Meeting, San Antonio, TX (2010).
7. United States Patent #6,255,118 "Method for Using an All Solid-State Fluorometer in Industrial Water Applications"
8. United States Patent #5,736,405 "Monitoring Boiler Internal Treatment with Fluorescent-tagged Polymers"
9. "Monitoring and Control in Boilers With All Solid-State Fluorometers", Martin Godfrey, Corrosion / 2002, Paper No. 02012, NACE International Conference, Houston, TX (2002).

For presentation at the AIChE 2014 Spring Meeting; New Orleans, LA; March 31- April 3, 2014

353051 Steam Traps: Audience Survey of Operation and Maintenance

Abstract:

All chemical plants have steam traps. However, steam trap operation and maintenance is not identical from one chemical plant to another chemical plant. Their design and operation is not proprietary since they are standard purchased items at all chemical plants. Therefore, this presentation will be conducted in an "open" fashion; i.e., the audience can and will be encouraged to participate freely in it.

The presenters will moderate an interactive discussion with the audience pertaining to the operation and maintenance of steam traps. This discussion will delve into steam trap management and operation on a company-side basis and, more likely, on a site-basis. The presenters will also lead a steam trap maintenance discussion. In this latter discussion, an attempt will be made to establish a "best practices" approach to steam trap maintenance.

This session, in general, and this presentation, in particular, will provide chemical plant engineers with pertinent information regarding steam traps.

PROSYS™ Maximize Your Plant.

Operator Errors and What Can be Done to Minimize

Global Congress on Process Safety
March 31, 2014

10TH GLOBAL CONGRESS ON PROCESS SAFETY

PROSYS™ Maximize Your Plant.™

10TH GLOBAL CONGRESS ON PROCESS SAFETY

Presenter

Tom Nolan

- **Graduated from Ohio University with B.S. in Chemical Engineering**
- **24 years experience in chemical process industry in a variety of roles**
- **MAIC and DFSS Six Sigma Black Belt**

PROSYS Maximize Your Plant™

10th GLOBAL CONGRESS ON PROCESS SAFETY

The Cost of Errors

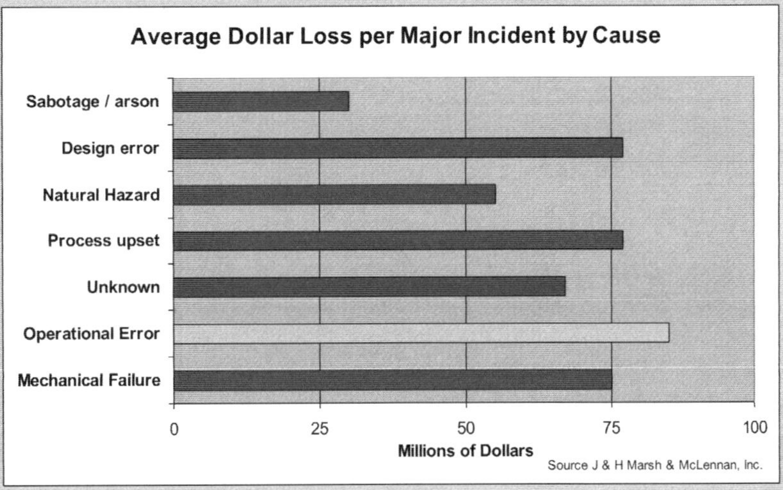

PROSYS Maximize Your Plant™

10th GLOBAL CONGRESS ON PROCESS SAFETY

The Cost of Operator Errors

- **ASM estimates total loss due to operator error is $8B per year**
- **Chemical Safety Topical Committee - average of one chemical incident per day - cost of over $2 million per incident to comply with requirements**
- **Errors cause 42% of unscheduled shutdowns**
- **70% of process incidents occur during start-up or shutdown**

PROSYS Maximize Your Plant™

10th Global Congress on Process Safety

Reasons for Errors

- **Lack of Skill**
- **Lack of Knowledge**
- **Carelessness**
- **System Design**
  - **Operator set up to make errors by inappropriate design – built in errors**
  - **Errors that are predictable are preventable by better design**

PROSYS Maximize Your Plant™

10th Global Congress on Process Safety

Focus of this Discussion

- **System Design**
  - **Alarm Management**
  - **Operator Graphics**
  - **Operator set up to make errors by inappropriate design – built in errors**
  - **Errors that are predictable are preventable by better design**

PROSYS Maximize Your Plant.™

10th Global Congress on Process Safety

Alarm Floods

In a number of industrial incidents, alarm floods were identified as a significant contributing cause to the incident...

As found by EEMUA in 1999 and CSB

Alarm Flood defined by ISA 18.2 as -
"10 or more annunciated alarms in any 10 minute period per operator"

PROSYS Maximize Your Plant.™

10th Global Congress on Process Safety

Why do Alarm Floods Occur?

One reason is not providing dynamic alarm management

- **Alarms need to indicate abnormal situations that require operator action**
- **Processes do not operate in one state**
- **What is normal vs abnormal changes with state**
- **Alarms are typically configured for run therefore many alarms are triggered upon a Change of State: Run to Shutdown**
- **Many of the alarms are not applicable or actionable for the new state impleading the operators ability to act quickly on what is important**

PROSYS Maximize Your Plant™

What is Impacted by Alarm Floods?

- **Product quality**
- **Operability or profitability of the process**
- **Loss of equipment**
- **Loss of containment – environmental releases**
- **Injury and loss of life in plant or community**

PROSYS Maximize Your Plant™

What Makes Alarm Floods so Dangerous?

Can be a problem for three reasons:

- **A deluge of alarms can cause critical alarms to be missed**
- **Floods can be a significant distraction when dealing with process upsets**
- **Can be an indicator of larger systemic safety issues**

PROSYS Maximize Your Plant™

10th Global Congress on Process Safety

Impacting Alarm Management Design

- Alarm rationalization is <u>not</u> a process to eliminate alarms - its about quality
- Good rationalization will add alarms when appropriate
- Typical (Static) rationalization is only for run mode
- Dynamic rationalization considers all plant modes

PROSYS Maximize Your Plant™

10th Global Congress on Process Safety

Dynamic Alarm Management

- Dynamic rationalization does everything a static rationalization does plus asks "when" for each alarm
- Requires dynamic software to make changes based upon operating mode of the plant
- Eliminates redundant alarms and lowers operator loading during transitions
- Only alarms what is abnormal and actionable for the given state

PROSYS Maximize Your Plant™

10th Global Congress on Process Safety

Actual Performance Metrics vs ISA 18.2

Type	Avg Alarm Rate/hr	Low/Hi of Avg Rate/hr	Peak Alarm Rate/hr	Low/Hi of Peak Rate/hr	% Time in Flood	Standing Alarms >24 hrs
Before Rationalization	30	Low=6.2 Hi=61	638	Low=152 Hi=2402	17.2%	9
ISA 18.2 Target Metrics	6	---	<60	---	<1%	<5
After Dynamic Alarm Management	2	Low=0.09 Hi=2.5	25	Low=13 Hi=42	0.25%	4

PROSYS Maximize Your Plant™

10th Global Congress on Process Safety

Results After Dynamic Alarm Management

Annunciated Alarms Per Hour (Monthly) **Area: G1-G5-G6** **June 2013**

Average Counts | Peak Counts | Total Counts

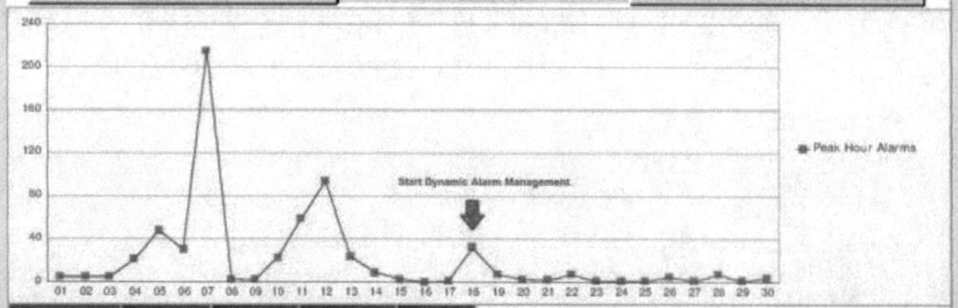

Day	Min Hour Alarms	Avg Hour Alarms	Peak Hour Alarms	Total Alarms
01-Jun-2013	0	0.3	5	6
02-Jun-2013	0	0.4	5	9
03-Jun-2013	0	0.5	5	13
04-Jun-2013	0	1.1	21	27
05-Jun-2013	0	13.3	48	319
06-Jun-2013	0	3.6	30	86
07-Jun-2013	0	11.9	215	285
08-Jun-2013	0	0.4	3	10
09-Jun-2013	0	0.3	2	7
10-Jun-2013	0	1.5	23	35
11-Jun-2013	0	4.7	59	112
12-Jun-2013	0	11.9	94	285
13-Jun-2013	0	3.8	24	90
14-Jun-2013	0	1.3	9	32
15-Jun-2013	0	0.3	3	8

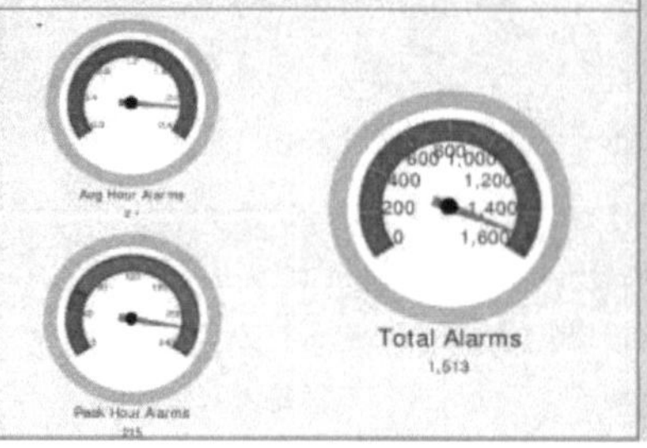

PROSYS™ Maximize Your Plant™

10th GLOBAL CONGRESS ON PROCESS SAFETY

Operator Graphics – The Risk of Changing Focus

- **Distraction of changing focus from process graphic to faceplate window**
- **Added workload from managing multiple open windows**
- **Increased probability of errors when changing values for un-intended tag with multiple faceplates open**
- **All of above become multiplied and more complex when process is transitioning from one state to another**

PROSYS™ Maximize Your Plant™

10th GLOBAL CONGRESS ON PROCESS SAFETY

Faceplate Operation

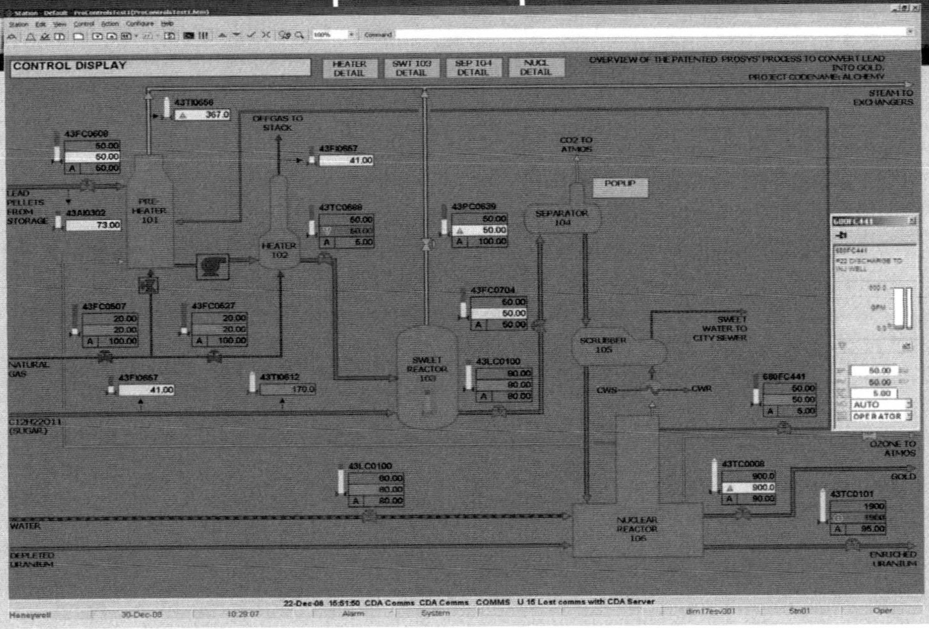

PROSYS Maximize Your Plant.™

Potential for Errors

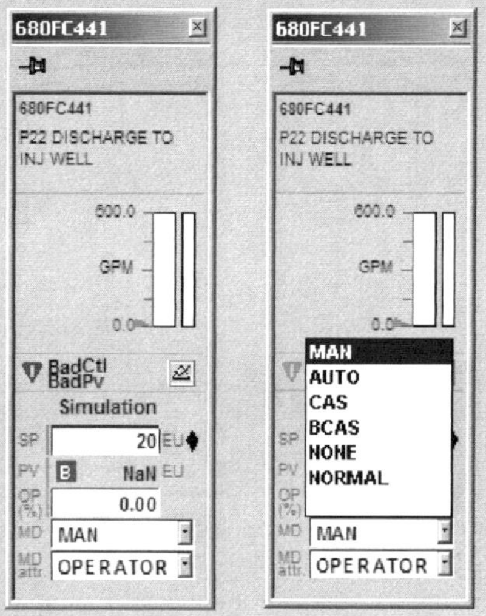

Faceplate Design -can introduce additional errors

- Mode drop down list covers SP, PV and OP values
- Operator may select wrong Mode if SP, PV and OP values are hidden from view
- Mode list offers more options than necessary

PROSYS Maximize Your Plant.™

Direct Entry Operation

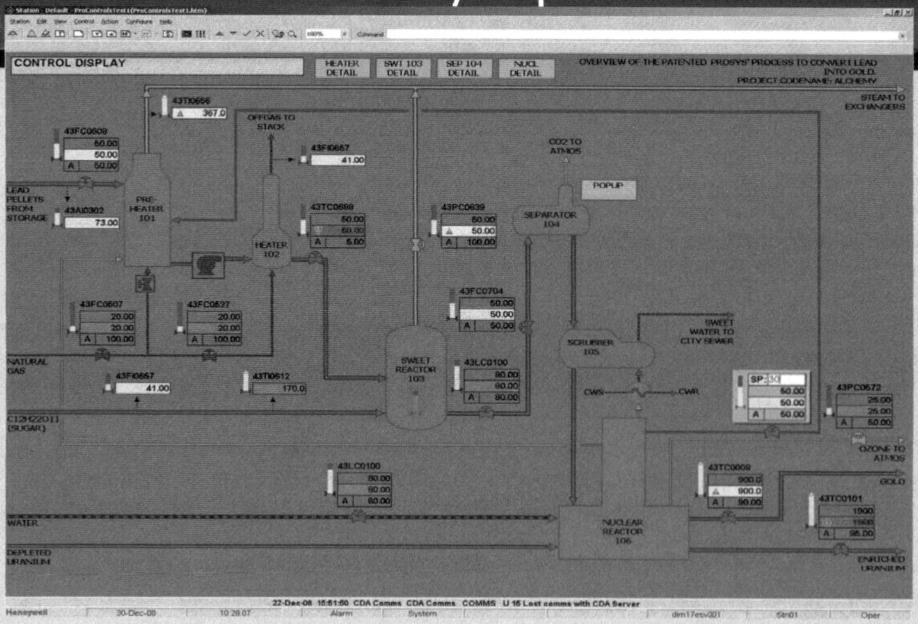

Direct Entry Fields

Values can be entered by:

- Select point
- Typing value via keypad
- Pressing enter

Key Factors:

- By maintaining focus of points after value change, it improves safety for input errors, multiple clicking functions and is more time efficient
- Easily recognizable operator enterable fields

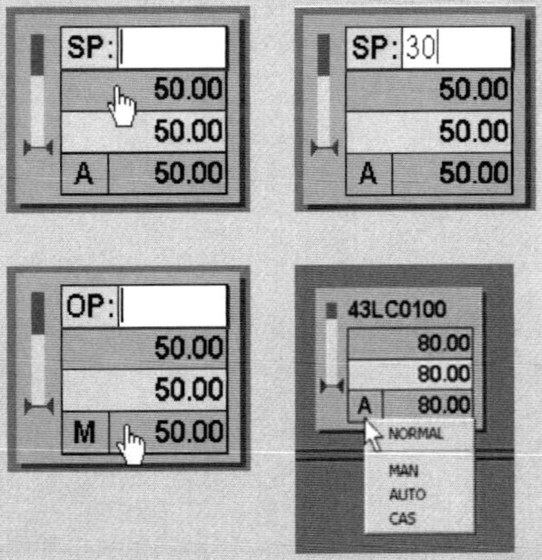

PROSYS Maximize Your Plant.™

Evaluation of Methods

Comparison of Faceplate to Direct Entry

- **Keystroke Level Analysis**
- **Operator Loading Analysis**
- **Risk Analysis**

PROSYS Maximize Your Plant™

Keystroke Level Model

Symbol	Time (s)	Description
K	0.28	Keystroke
P	1.1	Point to object
BB	0.2	Click on object
H	0.4	Home hands on keyboard or mouse
M	1.2+	Mental act or routine thinking

PROSYS Maximize Your Plant™

Estimate Execution Time Analysis

- **Evaluates execution time by an operator comparing faceplates to direct entry for SP/OP and Mode Changes**

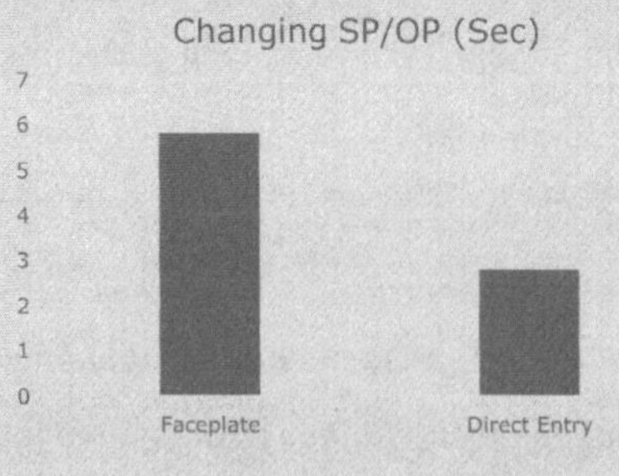

- **50 % Reduction**

PROSYS Maximize Your Plant™

10th GLOBAL CONGRESS ON PROCESS SAFETY

Estimated Operator Loading Analysis

- **Evaluates Mental and Physical Load on Operator**
  - Tasks that require very little thought are removed – such as (BB), (H)

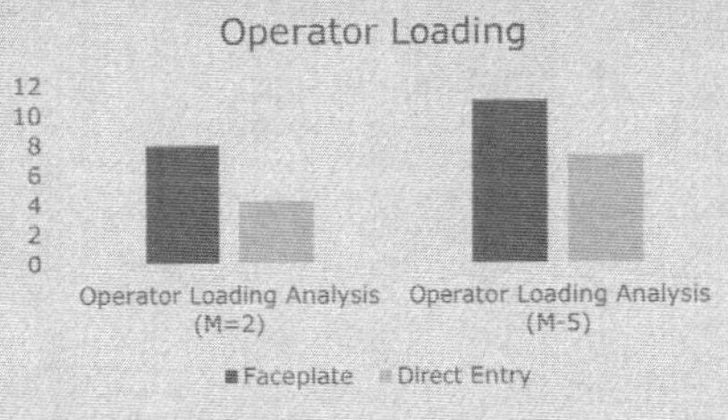

  - Mental and Physical tasks are weighted based on the amount of fatigue they cause – assigned a weight of M=2.0 to 5.0

PROSYS Maximize Your Plant™

10th GLOBAL CONGRESS ON PROCESS SAFETY

Risk Analysis

- **Identifies actions where operator entry errors can occur**

PROSYS Maximize Your Plant™

10th GLOBAL CONGRESS ON PROCESS SAFETY

Risk Analysis

Faceplate operation KLM for changing SP/OP	Potential Error
Move to shape (P)	Select wrong point
Click on shape (BB)	
Move to faceplate (P)	Point in faceplate is previous point
Click on SP/OP field (BB)	Change wrong parameter
Move hands to keyboard (H)	
Type in value and press enter (4K)	Mistype value and press enter
Move hands to mouse (H)	
Move to close faceplate (P)	
Click to close faceplate (BB)	
Potential error count	**4**

PROSYS Maximize Your Plant™

10th GLOBAL CONGRESS ON PROCESS SAFETY

Risk Analysis

Direct Entry operation KLM for changing SP/OP	Potential Error
Move to SP/OP (P)	Select wrong parameter or point
Click on SP/OP field (BB)	
Move hands to keyboard (H)	
Type in value and press enter (4K)	Mistype value and press enter
Potential error count	**2**

PROSYS Maximize Your Plant™

Direct Entry Vs Faceplate

Summary of Benefits

- **Execution Time SP/OP Changes - 51.5% less**
- **Execution Time for Mode Changes – 50% less**
- **Operator Loading – 33 % to 47% less**
- **Risk of Errors – 50% less**

PROSYS Maximize Your Plant™

Conclusion

- **Many industrial errors are a result of operators using control systems with flawed designs**
  - Poor design and performance by alarm management distracts operators and/or occludes critical alarms
  - Operator graphics with multiple open faceplates can cause a change intended for one controller to be entered into the faceplate of another
- **The two mechanisms listed above are predictable and therefore preventable through better design available today**

PROSYS™ Maximize Your Plant.
Questions?
Comments?
10TH GLOBAL CONGRESS ON PROCESS SAFETY

Problem Statement

visium KMS
Process Safety Management Solution

- Any organisation can improve by leveraging knowledge and lessons learned.
- Knowledge can be externally or hidden locally, in a global enterprise, where lessons learned are difficult to share beyond their initial impact.
- There is so much data, information and knowledge (opportunities) coming at your company that can make you stronger or make you look like your not paying attention.
- Not paying attention or managing the available knowledge, can sink your ship… nobody wants that!

Expert Insight Delivered

2

Osys
Part of the Rolls-Royce Group

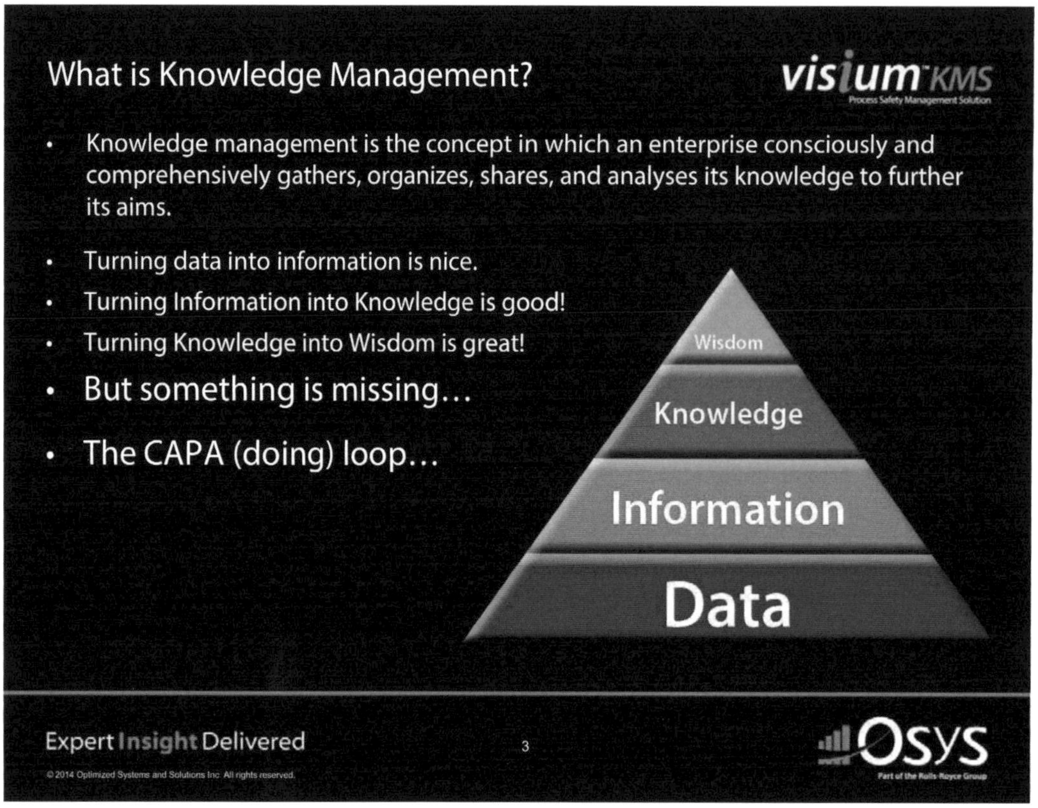
What is Knowledge Management?
visium KMS
Process Safety Management Solution
• Knowledge management is the concept in which an enterprise consciously and comprehensively gathers, organizes, shares, and analyses its knowledge to further its aims.
• Turning data into information is nice.
• Turning Information into Knowledge is good!
• Turning Knowledge into Wisdom is great!
• But something is missing…
• The CAPA (doing) loop…
Wisdom
Knowledge
Information
Data
Expert Insight Delivered
3
© 2014 Optimized Systems and Solutions Inc. All rights reserved.
OSyS
Part of the Rolls-Royce Group

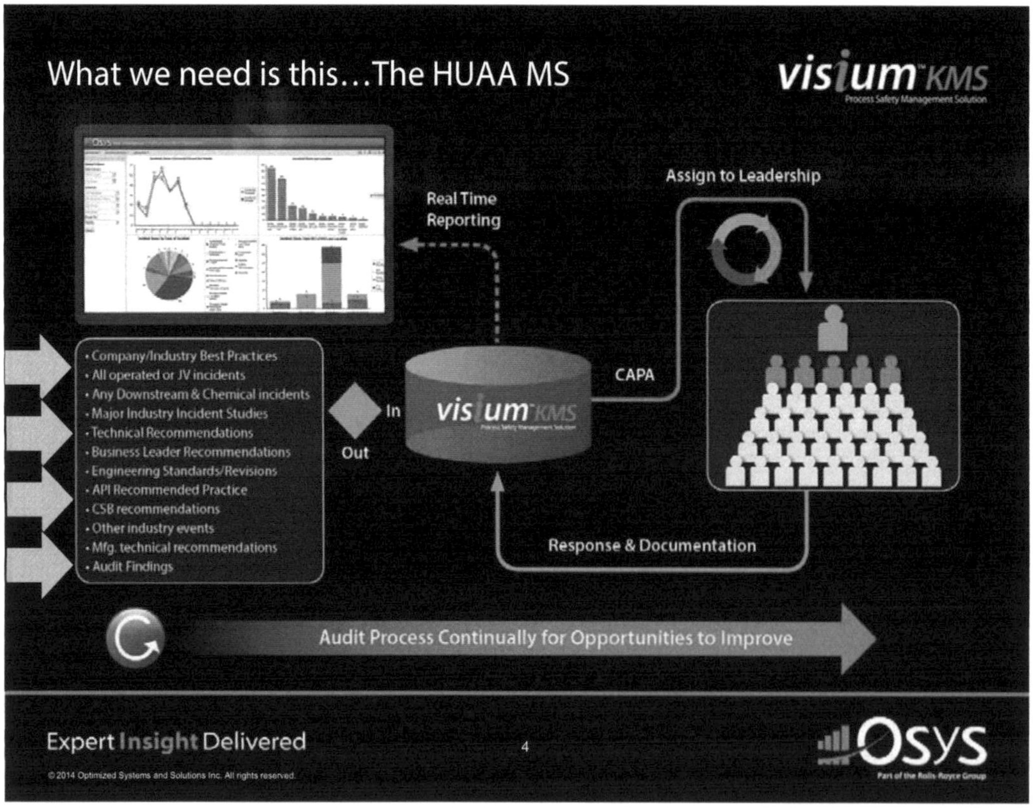
What we need is this…The HUAA MS
visium KMS
Process Safety Management Solution
Real Time Reporting
Assign to Leadership
• Company/Industry Best Practices
• All operated or JV incidents
• Any Downstream & Chemical incidents
• Major Industry Incident Studies
• Technical Recommendations
• Business Leader Recommendations
• Engineering Standards/Revisions
• API Recommended Practice
• CSB recommendations
• Other industry events
• Mfg. technical recommendations
• Audit Findings
In
Out
visium KMS
CAPA
Response & Documentation
Audit Process Continually for Opportunities to Improve
Expert Insight Delivered
4
© 2014 Optimized Systems and Solutions Inc. All rights reserved.
OSyS
Part of the Rolls-Royce Group

What is the HUAA Management System…

visium KMS
Process Safety Management Solution

HUAA – Heard, Understood, Acknowledged, Actioned.

1. Opportunities identified (heard) by your experts,
2. Captured by an electronic system,
3. Expertly Reviewed (understood),
4. Accepted or rejected (acknowledged) and
5. Assigned (actioned) to Sr. Managers for resolution and tracked to closure.

Expert Insight Delivered 5

OSYS
Part of the Rolls-Royce Group

HUAA – Management Systems - Detail

visium KMS
Process Safety Management Solution

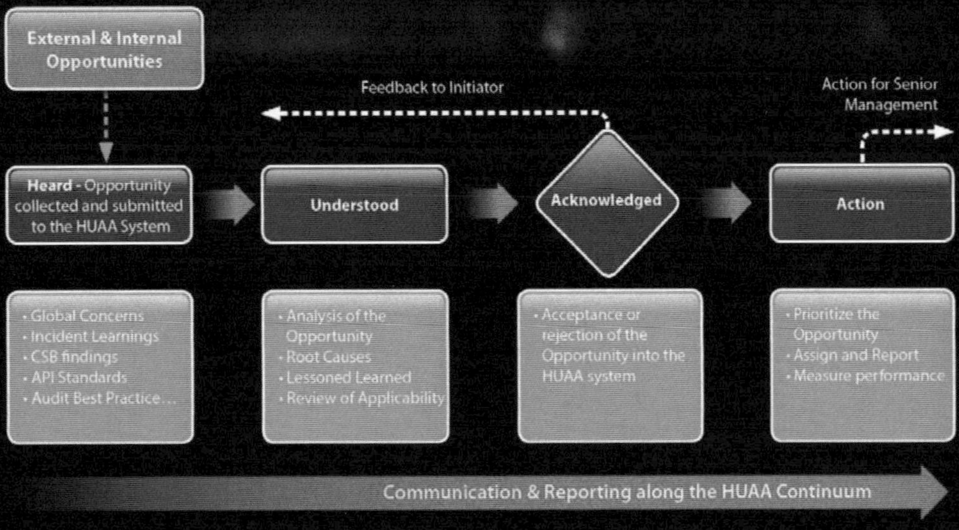

Expert Insight Delivered 6

OSYS
Part of the Rolls-Royce Group

HUAA Results

visium KMS
Process Safety Management Solution

- Efficient Collection - Important Opportunities are identified and collected from many sources (internal and external), and entered into the HUAA management system.
- Expert Vetting - These opportunities are vetted, parsed and resolved for inclusion or exclusion; documenting & communicating both.
- Accountable Assignment - Included opportunities are systematically assigned to Sr. Leaders and managed across the enterprise at SVP and C levels.
- Action - Improvement will be assured through the accountable actions of these leaders.
- Results - The organization will gain competitive advantage by active listening, learning from their experience, as well as the experience of others, and by taking ACTION!

Expert Insight Delivered 7 Osys

Part of the Rolls-Royce Group

Sharing is not enough to keep your ship afloat!

visium KMS
Process Safety Management Solution

Expert Insight Delivered 8 Osys

Part of the Rolls-Royce Group

352631 Autocad - Smart Grid – Agriculture - Architecture and the United States Government

Abstract:

The computational methods, process control, process design, process safety, public policy of the smart chip for agriculture industry, and architecture industry in the United States utilizing the computational methods of AutoCAD in the structural development stages of the smart grid for device/component/equipment and geographical development and public policy for grant funding.

352747 Production, Characterization and Catalytic Studies of Biobased Carbon Materials

Abstract:

Carbon-based nano materials were produced from wood chips, lignin and bio-char. The influence of reaction conditions on the production of nano-carbons was investigated. Carbon-based nanomaterials were characterized using scanning electron microscopy (SEM), transmission electron microscopy (TEM), high resolution transmission electron microscopy (HRTEM), and X-ray diffraction (XRD). The catalytic performance of these nanomaterials as syngas cleaning adsorbents and chemical conversion catalysts for bio-fuel processes has been evaluated.

347835 Uncertainty in Sour Gas Viscosities Estimation, What Is the Effect on Your Reservoir Inflow and Tubing Performance

Abstract:

In light of the need to find an applicable, consistent and promising correlation to accurately calculating sour gases viscosity under several reservoir conditions, this study aims to compare the existing correlations and other industry standard methods, check their accuracy and range of applicability for sour/acidic gases. Moreover, investigating the impact of uncertainties in estimating sour gas viscosities on reservoir flow performance relationship (IPR), reserves and resources estimations, tubing performance, pipeline and facilities design, the design of compressors and more importantly the need to find such correlation to be used in further advancements of reservoir simulation.

354052 Chromonic Nanocarriers for Chemotherapeutics: Size Distribution and Control Release Studies

Abstract:

Conventional chemotherapy fails to release the drug in the controlled manner; rather they release the drug instantly due to which re-administration is required as therapeutic index falls. Moreover, they target both the cancer cells and non-cancer cells which make the treatment of the cancer cells highly ineffective due to excessive toxicities. Recently, the nanoparticle mediated drug delivery for chemotherapeutics has gained much attention. The present study reports the use of a novel class of materials – *Chromonics* in cancer drug delivery. Folic acid as a chromonic material has been explored to develop nanocarrier for sustained release of chemotherapeutic drugs *in-vitro*. With the help of a biocompatible polymer as a tool, folate nano-domains were formed from self-assembled liquid crystalline folate solutions. These nano-domains were crosslinked to form a stable nanoparticle system. The effect of different factors like crosslinking agent, relative polymer and folate concentration, on the size distribution was investigated in detail. Based on the observations, we have reported that not only the synthesis but also the size of the developed nanoparticles can be controlled by varying the folate: HPMC weight ratios. The present study then explored the release of drugs from these nanocarriers. We have investigated the release of folic acid as a model drug from these nanoparticles as well as release of methotrexate and cytarabine as an anticancer drug for the period of 60 days. It was observed that release of drug was nanoparticle's size-dependent, release medium dependent, pH of the release medium dependent, crosslinking agent and its concentration dependent. The significant part of the current study is the synthesis of chromonic nanoparticles, how the size can be controlled, how controlled release can be obtained as well as how the release profiles can be modified depending on the dosage forms. Therefore, we can say that folate nanoparticles can be a potent nanocarrier for the sustained release of cancer drugs.

Double Containment Piping Systems: A Fail-Safe, Leak-Free Solution

IPEX
Committed to Excellence

Darin Johnson, P.Eng.
Applications Engineer
IPEX USA LLC

AIChE – 2014 Spring Meeting
New Orleans, LA

Monday, March 31, 2014

Why Double Containment Piping Systems?

IPEX
Committed to Excellence

- December 1988 – law change within the ***Resource, Conversation, and Recovery Act*** in the United States regarding underground piping.
- U.S. Environmental Protection Agency (EPA) requirement: Code of Federal Regulations (CFR), Title 40, Part 280.
  - Petroleum Products
  - Hazardous Substances (over 3200, CFR Title 40 Part 302.4)
  - Hazardous Wastes

Why Double Containment Piping Systems?

- Buried tank - all buried piping connected to it must be double contained.
- Above ground tank – associated buried piping consists of 10% or more of the tank volume, it also must be double contained.

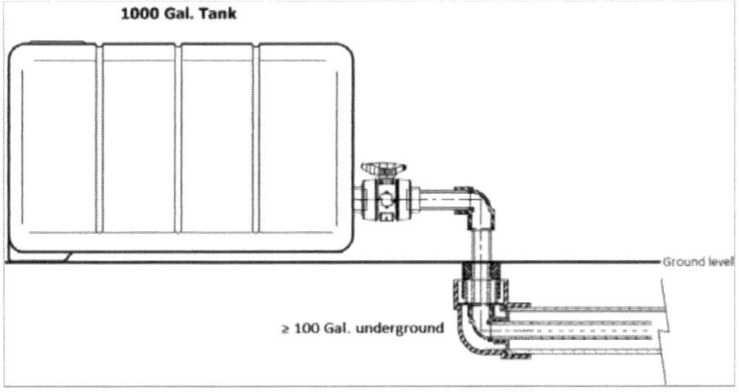

Why Double Containment Piping Systems?

- If the piping is less than 10% of the volume of the aboveground tank, and you choose not to double contain the buried piping, ***responsibility for any clean-up or remediation resulting from a release will still fall on the Client.***

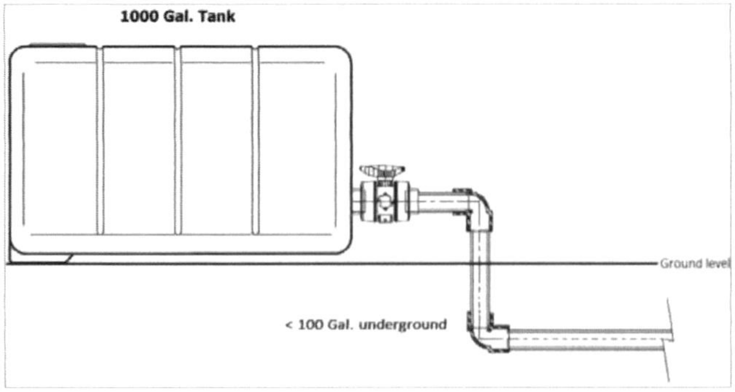

Why Double Containment Piping Systems?

- **Above Ground Piping Systems**
  - Different jurisdictions have to ability to require chemicals be double contained.

Nomenclature of Double Containment Piping

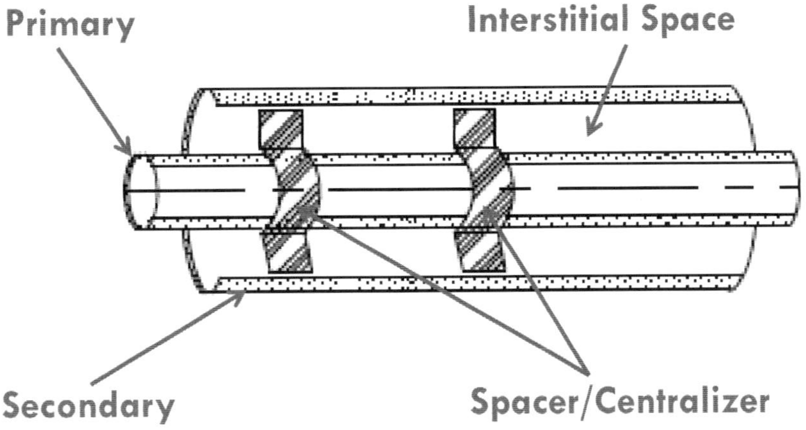

IPEX
Committed to Excellence

Double Containment Piping Systems: Material Combinations

SYSTEM Carrier	SYSTEM Containment	Carrier SIZE inches
PVC 80	PVC 80	1/2 - 12
PVC 80	PVC 40	1/2 - 12
PVC 40	PVC 40	1/2 - 12
PVC DWV	PVC DWV	1-1/2 - 12
PVC 80	CS SCH 10	1/2 - 12
CPVC 80	CPVC 80	1/2 - 8
CPVC 80	PVC 80	1/2 - 12
CPVC 80	PVC 40	1/2 - 12
CPVC 80	CS SCH 10	1/2 - 12
NAT PP 80	PVC 80	1/2 - 2
NAT PP 80	PVC 40	1/2 - 2
RED PVDF 80	CPVC 80	1/2 - 2
RED PVDF 80	PVC 80	1/2 - 2
NAT PVDF 80	PVC 80	1/2 - 2
ENFUSION PP	PVC DWV	1/2 - 6
S40 A53B CS BW	PVC 80	1/2 - 8
S40 A106B CS BW	PVC 80	1/2 - 8
S40 A53 BLK SW	FRP	1/2 - 2
S40 A53 BLK SW	PVC 80	1/2 - 2
S80 A53B CS SW	PVC 80	1/2 - 2
S10 T304L SS BW	PVC 80	1/2 - 8
S10 T304L SS BW	PVC 40	1/2 - 8
S10 T0304L SS BW	CPVC 80	1/2 - 8
S10 T304L SS BW	FRP	1/2 - 12
S10 T304L SS BW	S10 T304L	1/2 - 12
S40 T316L SS BW	PVC 80	1/2 - 8
S40 T316L SS BW	CPVC 80	1/2 - 8
S40 T316L SS BW	FRP	1/2 - 12
S40 T304L SS SW	PVC 80	1/2 - 2
S40 T304L SS SW	PVC 40	1/2 - 2
S40 T304L SS SW	CPVC 80	1/2 - 2
S40 T304L SS SW	FRP	1/2 - 2
S10 T316L SS SW	PVC 80	1/2 - 2
S10 T316L SS SW	CPVC 80	1/2 - 2

IPEX
Committed to Excellence

Double Containment Piping Systems: Design Considerations

- **Material Selection**
  - Primary Pipe
    - Chemical Compatibility
    - Pressure Capabilities
    - Thermal Properties
  - Secondary Pipe
    - Chemical Compatibility
    - Loading (Static & Live) for buried applications
    - Hanging requirements for above ground
    - Leak detection requirements
    - UV requirements

Double Containment Piping Systems: Design Considerations

- **Thermal Expansion & Contraction**
  - Two basic thermal relationships may be present:
    - When only one of the two systems experiences thermal expansion/contraction.
    - When both systems experience different magnitudes of thermal expansion/contraction.
  - The Floating System
  - The Restrained System

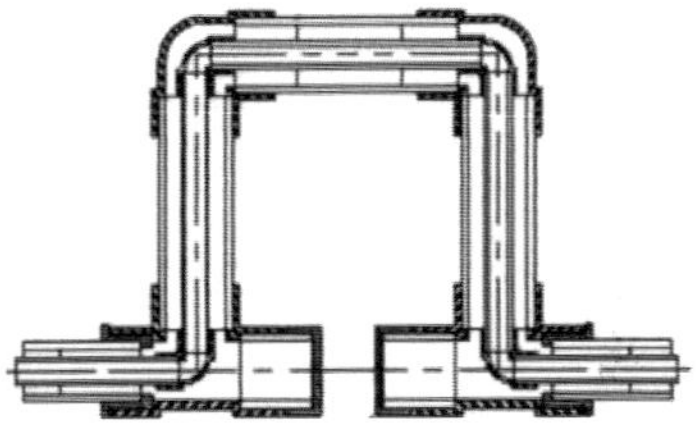

Double Containment Piping Systems: Leak Detection Requirements

- EPA requirement: (CFR), Title 40, Part 280
  - Pressurized systems require electronic leak detection that must be checked once per year.
  - Drainage systems (gravity flow) require visual leak detection with monthly manual inspection for a media leak.

Double Containment Piping Systems: Leak Detection Requirements

- The following forms of leak detection are acceptable as per regulations

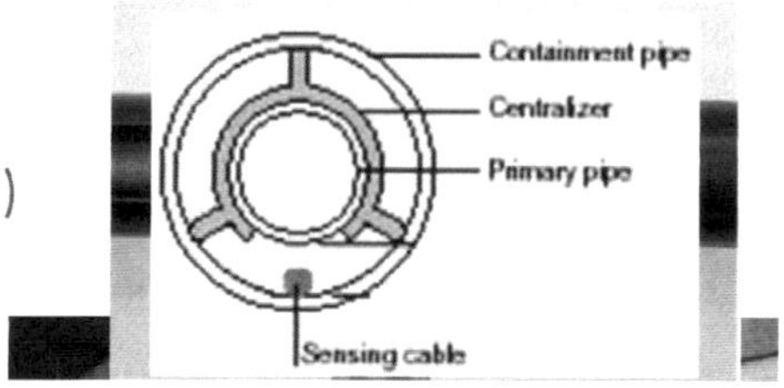

  - Visual (drainage applications only)
  - Electronic low point probes
  - Continuous cable

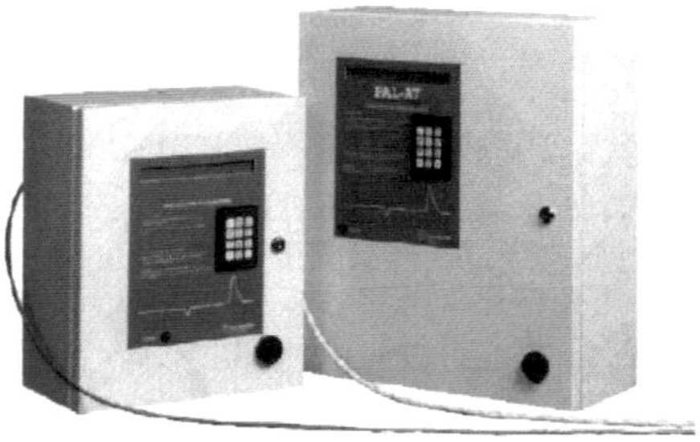

Double Containment Piping Systems: Applications

- Underground Applications
- WTP & WWTP
- Oil & Gas
- Hospitals
- Museums
- Food and Beverage
- Pharmaceuticals
- Laboratories
- Data Centers

Questions?

Thank you!

DTP Process

"Now Ready to License"

JGC Corporation
Mitsuo Morita

2014

JGC CORPORATION | MITSUBISHI CHEMICAL

What is DTP Process ?

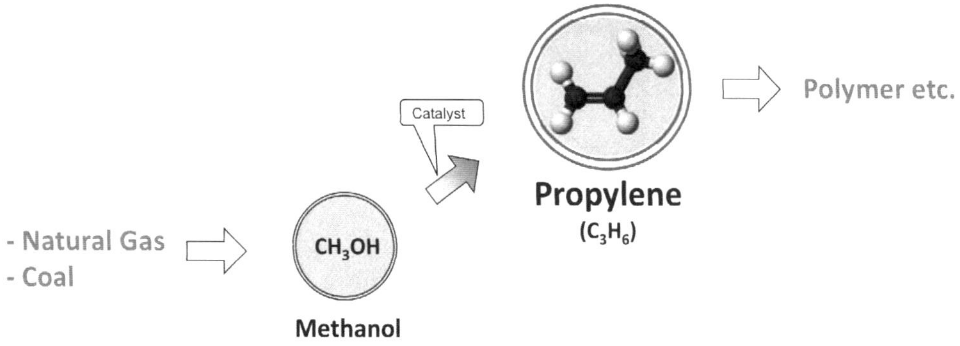

-On-Purpose Propylene Production from Methanol

JGC CORPORATION | - 2 - | MITSUBISHI CHEMICAL

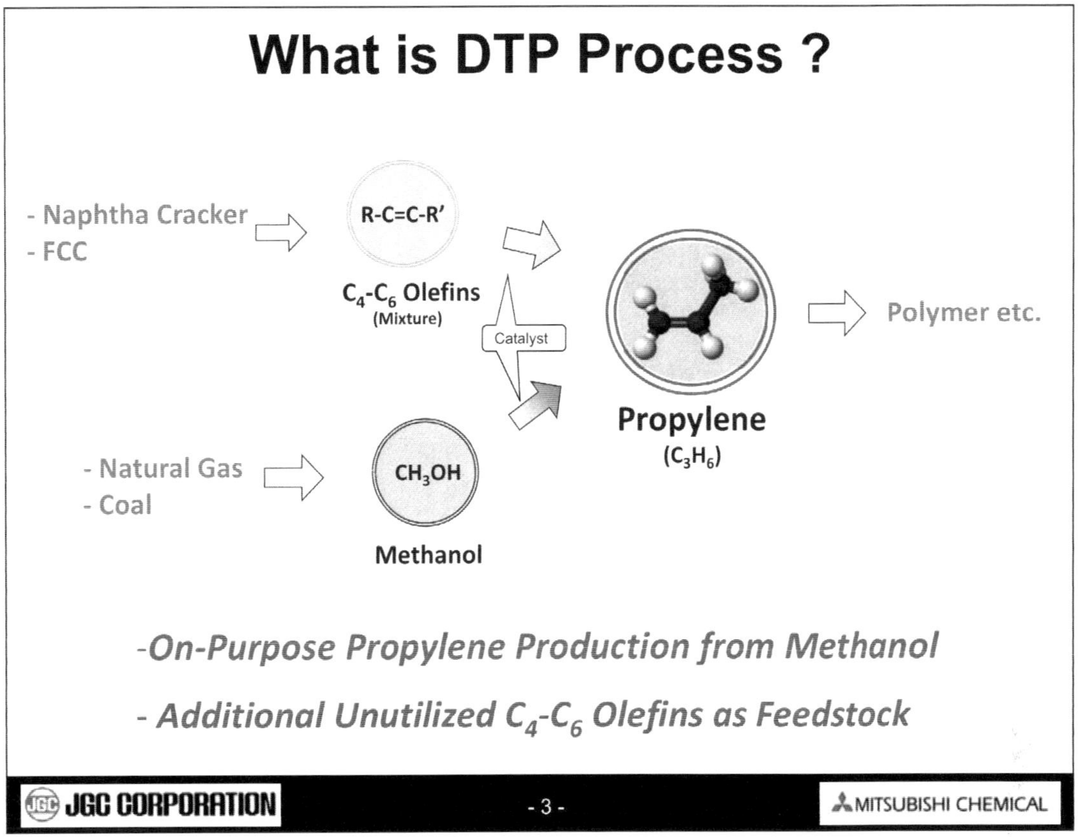
What is DTP Process ?
- Naphtha Cracker
- FCC
R-C=C-R'
C_4-C_6 Olefins
(Mixture)
Catalyst
Polymer etc.
Propylene
(C_3H_6)
- Natural Gas
- Coal
CH_3OH
Methanol
-On-Purpose Propylene Production from Methanol
- Additional Unutilized C_4-C_6 Olefins as Feedstock
JGC CORPORATION
- 3 -
MITSUBISHI CHEMICAL

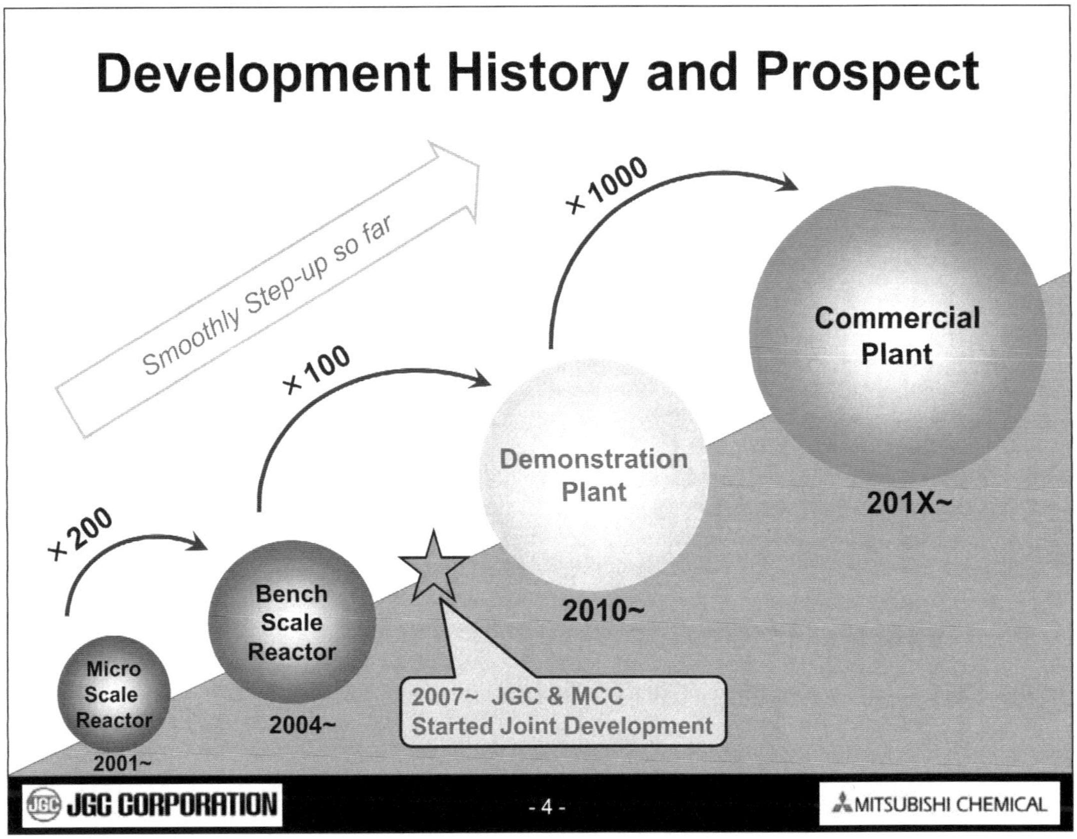
Development History and Prospect
Smoothly Step-up so far
×1000
×100
×200
Commercial Plant
201X~
Demonstration Plant
2010~
Bench Scale Reactor
2004~
Micro Scale Reactor
2001~
2007~ JGC & MCC Started Joint Development
JGC CORPORATION
- 4 -
MITSUBISHI CHEMICAL

Demonstration Plant Overview

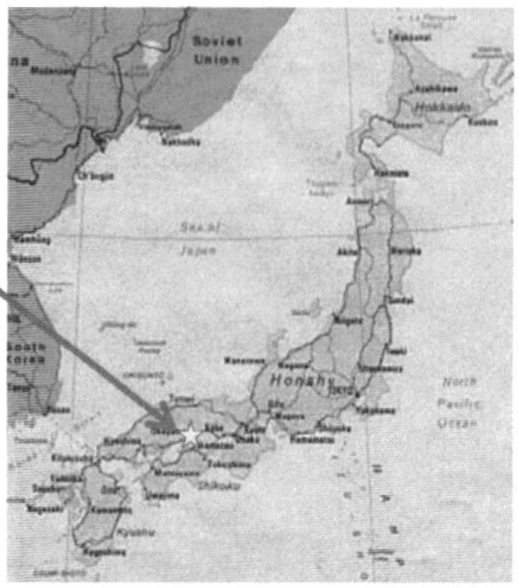

- **Located in Mitsubishi Chemical Mizushima Plant**
- **Successfully Start-up in August 2010**
- **Continuous Operation for Two Years**

Process Flow (Standalone Case)

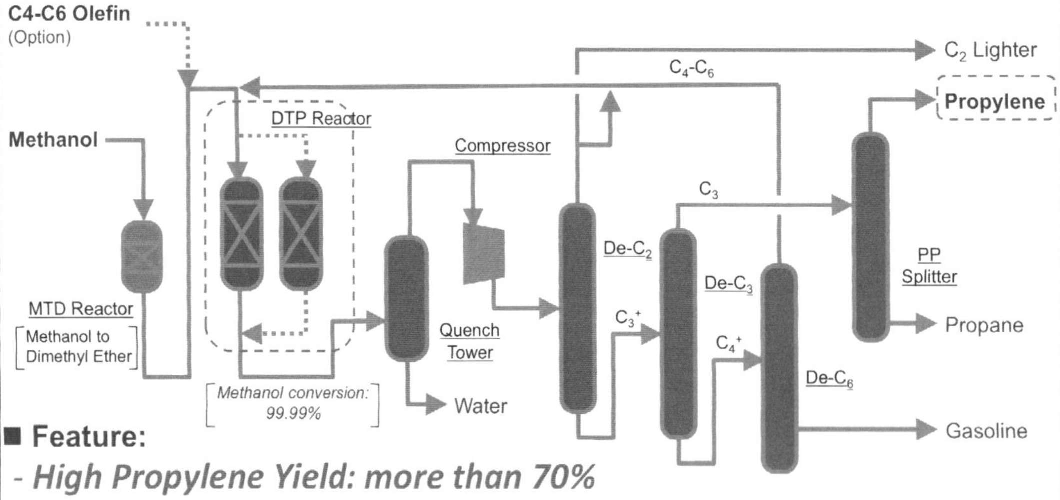

■ **Feature:**

- *High Propylene Yield: more than 70%*
- *Applied Existing Proven Technologies*

Process Flow (Integration Case)

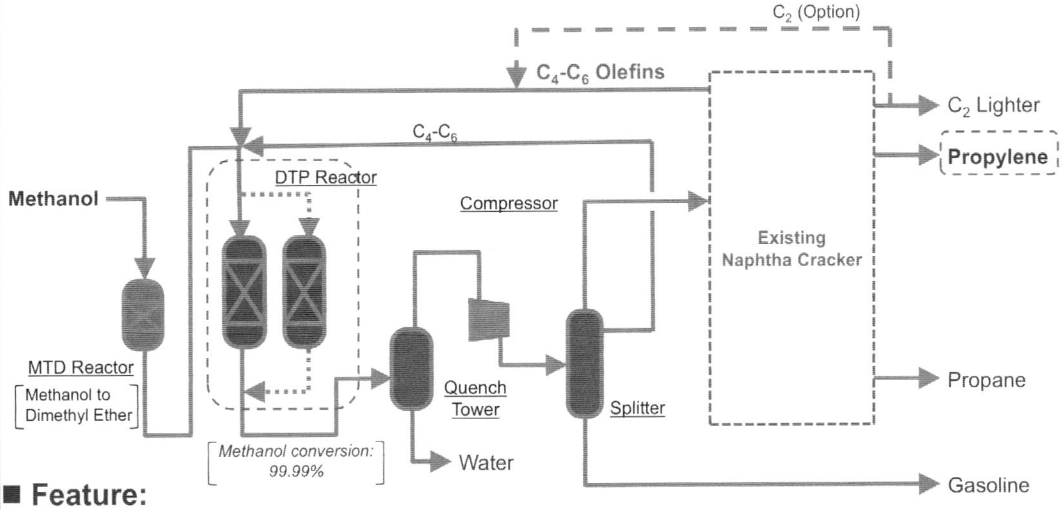

■ **Feature:**

- *High Propylene Yield: more than 70%*
- *Applied Existing Proven Technologies*
- *Low Investment Cost in case of Integration with Naphtha Cracker*

Application

(1) Naphtha Cracker Integration Case (Propylene Capacity: 20KTA – 200KTA)

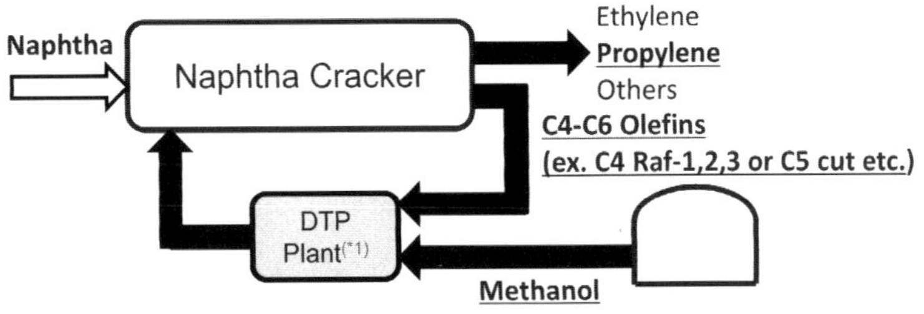

(*1) Utilizing facilities in Naphtha Cracker.

(2) Standalone Case (Propylene Capacity: 250KTA – 500KTA)

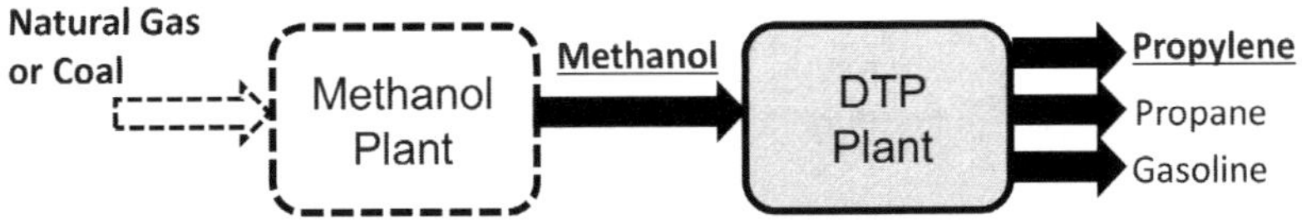

- *Two Cases available depending on Client requirements !!*

DTP is…

- *High Yield Propylene Production Process*
- *Flexible Feed Methanol to Olefin Process*
- *Process that increase P/E Ratio for Cracker*

Thank you very much !!

Contact:
Mitsuo Morita
Market Development Dept.
Global Marketing Division
2-3-1, Minato Mirai, Nishi-ku,
Yokohama 220-6001, Japan

Phone: 81-45-682-8371
FAX: 81-45-682-8183
E-Mail: morita.mitsuo@jgc.co.jp

352533 Self-Cleaning "Bernoulli" Type Filters Used in Onshore and Offshore Applications

Abstract:

Georg Schuenemann GmbH (SAB-BREMEN) has developed a new Filter Type F-480, dubbed the "Compact Giant" for applications in onshore water intake and cooling water systems, sea water desalination plants, and in offshore projects, such as FPSO's, Drillships, and Production Platforms.

This paper will describe the essential design features of the filters, based on the "Bernoulli principle" which result in a fully automated self-cleaning filter system, and choice of materials in a sea water service with high corrosion rates.

A case study will be presented in which stringent project requirements such as substantial flow rates, adaptibility to space limitations and pipe connections had to be observed. Reliability of the system, extended run times without the need for frequent replacement of filter elements, as well as ease of monitoring and operator involvement have been achieved.

352653 Effects of Support on Sulfur Tolerance and Regeneration of Pt Catalysts Measured By Ethylene Hydrogenation and EXAFS

Abstract:

We investigated the effect of support on sulfur tolerance and regenerability under reducing environments by measuring the rate of ethylene hydrogenation, combined with hydrogen chemisorption, and extended x-ray absorption fine structure (EXAFS). Alumina and titania (P25) supports were used to prepare Pt catalysts at 1 wt.%. The catalysts were tested after treatment in H_2S/H_2 at 250 °C followed by regeneration treatments in H_2 at 250 °C, 350 °C and 450 °C. The sulfided catalysts exhibited a 20-27 times decrease in the rate of ethylene hydrogenation and a 4-6 times drop in the Pt surface area. Regenerations up to 450 °C were unable to remove all the sulfur, as shown by the presence of Pt-S bonds by EXAFS. However, a partial recovery of the hydrogenation rate per mole of Pt was observed on sulfided Pt/Al_2O_3 after reduction at 450 °C, while the induction of strong metal support interactions (SMSI) at reduction temperature above 350 °C was observed on Pt/P25, regardless of the presence of sulfur. For Pt/P25, the reversal of the SMSI state together with sulfur removal by mild oxidation suggests that sequential reduction/oxidation treatments may be more effective in restoring the S-free state of TiO_2-supported catalysts.

352799 Improved Operational Efficiency and Reliability through Insulation Materials Selection

Abstract:

This paper will explore the implications of insulation materials selection on the energy efficiency and reliability for Ethylene production units. Considerations will include insulation properties, structural support, installation methods, installation and maintenance costs.

BOILER MACT COMPLIANCE: MERCURY IN FUEL GAS

PATRICK LAINE, WES YONGE, & VIC VICKERY

PEI, HOUSTON, TX 77043

Patrick Laine, Ph.D.
Chief Scientist
Houston, TX
850-866-6686
plaine@pei-tx.com

Wes Yonge, PE
Chief Engineer
Houston, TX
713-806-8396
wyonge@pei-tx.com

James 'Vic' Vickery
Chief Technology Officer
Houston, TX
281-536-0899
vvickery@pei-tx.com

1 Introduction & Objectives

For most ethylene producers the primary objective of this sampling and analysis effort will be to verify and "establish a mercury fuel specification through which gas-fired boilers that use a fuel other than natural gas or refinery gas may be considered Gas 1 unit" as identified in 40 CFR Part 63.7521.(f).

Measurement and Monitoring Requirements: Decision Rules

- ❑ If Hg concentration ≤ 50% of Gas 1 specification* then no further testing required
- ❑ If Hg concentration > 50% and ≤ 75% of the specification* then semi-annual sampling required
- ❑ If Hg concentration > 75% of the specification* then monthly sampling required

*Specification is **40 µg/Sm³**

The goal will be to execute a sampling and analysis campaign that will produce compliance quality data. The key to producing and validating compliance quality data is a well-designed and executed QA/QC protocol. At a minimum, a performance based QA/QC protocol is recommended that includes the following components:

- Sorbent Trap Breakthrough Bed:
  - ❑ Sorbent traps consisting of a minimum of two sorbent beds in series so that the trailing bed can be analyzed to verify that no breakthrough occurred from the leading bed.
- Duplicate Samples:
  - ❑ The sampling system should be configured with two trains to accommodate the collection of simultaneous duplicate sorbent trap samples, in parallel, providing a means of verifying precision in the sample collection process.
- NIST (National Institute of Standards and Technology) Traceable Spike:
  - ❑ Sorbent traps with a third bed (downstream) that is spiked with a NIST traceable spiked mass of mercury serves as a matrix spike and provides a means of verifying that components in the matrix (the fuel gas) are not affecting the trapping efficiency of the sorbent trap.

2 Methods

Sorbent Trap Mercury Sampling Methods Comparison

Method	Description	Method Type: S = Sampling A = Analysis	Evaluation	Sample Collection Phase QA/QC
ISO 6978	Hg in Natural Gas by Thermal Desorption (CVAFS)	S & A	Method calls for low sample volumes can lead to non representative results.	No
EPA Method 30B	Determination of Gaseous Hg Emissions from Coal Fired Combustion Sources using Carbon Sorbent Traps	S	Requires modification (Modified M30B)	Yes
ASTM D6350	Hg S&A in Natural Gas (AFS)	S & A	Method calls for short term samples (i.e., low sample volume) which leads to poor accuracy.	No
ASTM D5954	Hg S&A in Natural Gas (AAS)	S & A	AAS is less sensitive than AFS.	No
EPA Method 1631	Measurement of Hg in Water	A	No	Yes

3 EPA Method 30B QA/QC Performance Criteria

QA/QC Test or Specification	Acceptance Criteria	Frequency
Analyzer calibration	Correlation Coefficient: 0.999	Daily
Transfer efficiency of analyzer	%Transfer Efficiency: 90%–110%	Once per sorbent trap
Initial calibration blank	< 0.5 ng	Once before each calibration
Sorbent Trap Section B breakthrough	≤ 10% of Section A Hg mass, for Hg concentrations > 1µg/dscm ≤ 20% of Section A Hg mass, for Hg concentrations ≤ 1µg/dscm	At each sample point
Spike Recovery Test	Average recovery between 75% and 125 % for Hg_0	Once per sample point
Paired Sorbent Traps Agreement	≤ 10% RD for Hg concentration > 1µg/dscm ≤ 20% RD or 0.2 µg/dscm absolute difference for Hg concentrations ≤ 1µg/dscm	At each sample point

4 Equipment

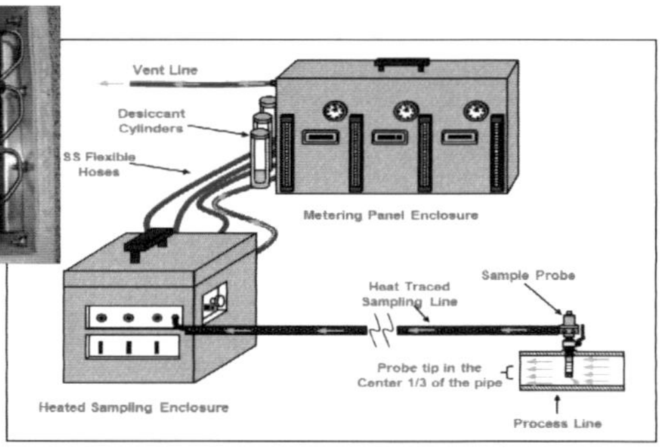

Flow

	Primary Bed	Breakthrough Bed	Spiked Bed

QA/QC Acceptance Criteria:	99% of Total Collected Hg	≤ 10% of Primary Mass	Recovery between 75% and 125%

Gold Trap:

5 Conclusions/PEI Recomendations

- ❑ There are five methods accepted by the Boiler MACT rule. PEI has extensive experience and expertise employing **all five** of these methods.
- ❑ Operators can achieve compliance without submitting a Fuel Analysis Plan by using ASTM D6350, however, we have noticed variable results with this method.
- ❑ The **PEI recommendation** for obtaining the most accurate compliance quality data is to use Modified EPA Method 30B in conjunction with EPA 1631 where traps should be analyzed in a certified ISO 17025 laboratory. The recommended pathway includes a rigorous, performance based QA/QC protocol that validates accuracy and precision of all sample collection procedures.

351376 Young Professional Simulation Tutorial

Abstract:

18C00 Young Professional Simulation Tutorial

Authors: Naomi Hua and Mike Donahue, Invensys

Process simulation is a complex skill to master. Even with years of experience, a typical user is constantly facing new problems that need to be addressed. There are many issues that can lead to unexpected and unnecessary errors with the use of process simulation tools. Regardless of what simulation program you are using, these issues are common from one program to another. In this presentation, targeted for the young professional, we will discuss the "Most Common Pitfalls in Process Simulation". For each topic, we will discuss various techniques and methods these troublesome problems and prevent their occurrence.

Topics to be discussed include:

1. Erroneous data entry

2. Simulation defaults

3. Bad plant data

4. Thermodynamic method selection

5. Simulation tolerances and convergence
6. Recycle identification and handling
7. Model complexity
8. Difficult or impossible specifications
9. Assay pseudo-component development
10. Assay property blending

1. Erroneous data entry is the most common mistake made when building a process simulation model. This presentation will cover methods that can help ensure proper data entry in your process simulation model, such as checking the data you entered, particularly the units of measure, using "alternate forms" of data input such as copy/paste and data import, and reviewing all warning and error messages.

2. Defaults are methods, values, and conditions that are automatically selected for you by the program when you begin a new simulation model. These defaults will affect the results calculated by your model and every simulation program has them. Most defaults within simulation programs fall under categories such as convergence criteria, thermodynamic methods, tolerance and units of measure. The problem with defaults is unintended results from not fully understanding the effects a particular default has on the model. Take the time to understand the defaults of the program and how these defaults will affect the results. If you do not agree with the defaults selected by the program, most simulation programs will allow the user to change the default.

3. Plant data changes over time. The feed information and process conditions of the unit operations may be different. Not only does plant data change over time, but measurement information is often very inaccurate. This could be due to faulty or poorly calibrated measurement equipment, leaks in lines, sensor failures, and other issues. Because of these changes and inaccuracies over time, you always have to validate the information that is entered in your model to ensure that you are obtaining the proper results. To do so, you can utilize programs that can help with data validation and reconciliation like SimSci's ROMeo online optimization program. With it, you can connect your model to your data historian to constantly import current data measurements into your model and then reconcile these measurements to determine when problems arise within your plant and where those problems are occurring. You will also need to use your engineering judgment to determine if the data is valid. More details will be provided during the presentation.

4. The proper selection of thermodynamic methods for your simulation is the most important decision that you make when modeling a process. The method you select could have a drastic effect upon your results. Often, any number of thermodynamic methods is suitable for a given application. It is up to the user to determine the best method to use based upon knowledge, personal experience, and best judgment. Some important factors that need to be considered are the components being used and the operating conditions.

Oftentimes, the selected thermodynamic method still needs to have additional inputs such as binary interaction parameters to simulate the process accurately.

5. Many of the calculations in process simulation are iterative. When these calculations iterate, the program compares the results calculated from one iteration to the results calculated in the next iteration in order to determine

whether it is converged to a solution or needs to continue calculating. This method of comparing results is where tolerances come into play and simulation programs use tolerances to determine when a solution is reached. One of the major mistakes when modeling a process is not understanding or using the correct tolerance values. A looser tolerance may make the file easier and quicker to converge, but it also reduces the accuracy of the model. A tighter tolerance may be more accurate, but might make it impossible for the model to solve or increase the convergence time exponentially. A common practice is that when you have reached a converged solution with your model, tighten the tolerances within the model and solve it again. If you notice that your answers have changed significantly, you need to continue tightening the tolerance values until your results change little from one iteration to another. You will then know that you have reached an accurate solution for your model.

6. Recycles in process simulation occur when you return to a previously calculated unit operation to recalculate the unit due to a change made elsewhere in the model. They are very common and can often be difficult to converge. Some suggestions on troubleshooting recycle convergence include initially solving your model without the recycles to see how it behaves, providing initial estimates for recycle streams, using tighter tolerances on unit operations within recycle loops than the total recycle tolerances themselves, and ensuring that all components have a way in and out of a recycle by performing mass balances and using purge or makeup streams where necessary.

7. The general rule of thumb when modeling is to start simple and add complexity as you go. Each time you add a few streams or unit operations converge and save the model before moving on. Make sure you justify any additions that you make to the model. You can most likely model your system without including every unit from your PFD in the simulation. When modeling, think about the goal of your simulation and not about literally translating every real-world piece of equipment into the simulation model. Make the judgment call on what is most important and what the purpose is of every unit you add to your simulation. For example, if your real-life process shows a heat exchanger, a pump and a splitter, but all that is necessary is that the temperature, pressure, and separation are achieved properly, consider using a flash drum to accomplish all three.

8. Specifications within process simulation are everywhere and every stream and unit operation you place in your model contains specifications. In its simplest form, a specification is a constraint that you place on the model that the model needs to meet in order to successfully converge. For streams, these are usually temperatures, pressures, flowrates and compositions. For unit operations, they range from process conditions, like temperatures and pressures, to product conditions, like an overhead composition in a column. There are some pitfalls, however, that you can run into with specifications such as azeotropes in columns, high purity specifications, and the possibility of multiple solutions.

9. Petroleum assays are sets of distillation data that are used to characterize all of the components that reside in a given crude sample. When you enter distillation data into a process simulator, the program converts it into components that it can use when modeling your process. The components that are generated are called pseudo-components, and each pseudo-component represents all of the unknown components that would be within a given boiling point range in the crude. Some of the things you will want to ensure when properly developing your assay components are did you use the correct distillation curve basis, did you enter the data correctly, did you use enough cuts and are you using the correct characterization methods .

10. When you have more than 1 assay in a simulation model, you need to be concerned with assay blending and how it will affect the results of your file. Assay blending is the mixing of 2 or more groups of components (i.e. ¨C sets of cuts/pseudo-components) in order to calculate averaged component properties for cuts that are in the same boiling point range as one another. Once these new averaged component properties are determined, the program will use these averages from this point on in the simulation model any time it needs to access component

properties for a given pseudo-component. All component properties are blended, including normal boiling points, specific gravity, molecular weight and refinery inspection properties, and usually the blended component properties are calculated using a weighted average

These topics are based on experience and feedback from end-users to technical support representatives for process simulators. They often represent the most commonly asked questions or areas that most users find to be problematic. While the specific examples highlighted in each topic come from SimSci software, the intent of this tutorial presentation is to be broad-based and is particularly applicable to young professionals seeking to develop their experience with simulation.

Polymeric Sulfonic Acid Composites for Fuel Cell Applications: Effects of Casting Solvent and Sonication Time on Membrane Performance

*Jimoh K. Adewole[1], Abdullah S. Sultan[1, *], Amir Al-Ahmed[2], and S. M. Javid Zaidi[3]*

[1]Center for Petroleum & Minerals, King Fahd University of Petroleum & Minerals, Dhahran 31261, Saudi Arabia

[2]Center of Research Excellence in Renewable Energy, King Fahd University of Petroleum & Minerals, Dhahran 31261, Saudi Arabia

[3]School of Chemical Engineering, The University of Queensland, Australia

Abstract

Polymer blends comprising sulfonated poly (ether ether ketone) (SPEEK) and 3wt% (2-acrylamido-2-methyl-1-propanesulfonic acid) (PAMPS) were prepared by solution casting method. Two solvents (N,N-Dimethylacetamide (DMAc) and N-Methyl-2-pyrrolidone (NMP)) were uesd separately. Residual solvents in the membrane samples after drying were quantified using thermo gravimetric analysis. Effects of periods of sonication (0 to 90 minutes) and types of solvent on the membrane performance were evaluated interms of morphology and water uptake. An attempt was made to correlate membrane performance with the time of sonication in order to select the most suitable solvent and membrane casting conditions for subsequent membrane preparations. For both solvents, the water uptake was observed to first decrease at 10 minutes of sonication and then increased as the sonication time was increased. The minimum water uptake recorded for DMAc samples was 19.23% at 20 minutes sonication time while that of NMP sample was 24.24% at 10 minutes. However membrane morphology was found to deteriorate with increase in sonication time. No correlation was obtained between the membrane properties and the sonication time. In general, membrane performance was better for membranes samples using DMAc than NMP solvent. Therefore, DMAc was selected as the casting solvent for membrane preparation on which further studies will be carried out.

Keywords: morphology, water uptake, residual solvent, thermogravimetric analysis

Introduction

Water soluble polymers have received academic and industrial interests due to their potential applications in fuel cells and other electrochemical devices (Ye et al. 2012). One of the water soluble polymers that have attracted publication interest is poly (2-acrylamido-2-methyl-1-propanesulfonic acid) (PAMPS) (Florjańczyk et al. 2003). Generally, some of the major performance criteria for fuel cell materials include proton conductivity, water and methanol uptake, good mechanical properties, chemical stability, thermal stability and methanol permeability (Gil et al. 2004; Swier et al. 2005; Ye et al. 2012). PAMPS was reported to posses higher proton conductivity than partially hydrated Nafion. In addition, it has been found to be more tolerant to water fluctuation in high temperature operations. Some of the major drawbacks in the use of PAMPS are its poor mechanical properties, thermal stability, water and methanol uptake and solubility in many solvents (Florjańczyk et al. 2003). One of the approaches that have been proposed to solve these problems is development of polymer composites.

This research paper investigates the synergistic effect of combining water soluble PAMPS with sulfonated poly (ether ether ketone) (SPEEK) which is non - water soluble polymer to prepare composite membranes for application in fuel cells. Specifically, preliminary results on the performnance of the composite were presented. Performance were evaluated based on the composite morphology and water uptake. Effects of sonication time and types of casting solvent on membrane properties were also evaluated.

Material and Method

Material

Sulfonated poly (ether ether ketone) (SPEEK) purcahsed from FuMA-Tech, Germany, and poly (2-acrylamido-2-methyl-1-propanesulfonic acid) (PAMPS) purchased from Aldrich Chemical Co., are the two major polymers used in this research. The SPEEK has a DS of 0.533 and solubility of 99% in DMAc. SPEEK is derived by sulfonation of Poly (ether ether ketone) (PEEK). PEEK is a commercially successful polymer due to its excellent thermal stability and solvent resistance (Mohr et al. 1991). For sometimes, further application of PEEK have been stalled due this solubility issues. By sulfonation, PEEK can be made soluble in most commercial solvents. Importantly, the solubility of SPEEK also depends on its degree of sulfonation (DS). Since, this work involves the use of two polymers, a solvent which will disolve both the PAMPS and SPEEK must be selected. Literature review showed that PAMPS dissolves in N,N-Dimethylformamide (DMF), N,N-Dimethylacetamide (DMAc), propylene carbonate (PC), and water (Florjańczyk et al. 2003). Initial solubility studies that was carried out in our laboratory was done for N,N-Dimethylacetamide (DMAc) and N-Methyl-2-pyrrolidone (NMP). The choice of NMP was due its unique solvent properties and its structtural distiction from DMF and DMAc (Figure 1). Moreover, no literature was found where NMP was used as casting solvent for PAMPS. Thus, DMAc and NMP were used as solvents in this work. All materials were used as received from the manufacturers.

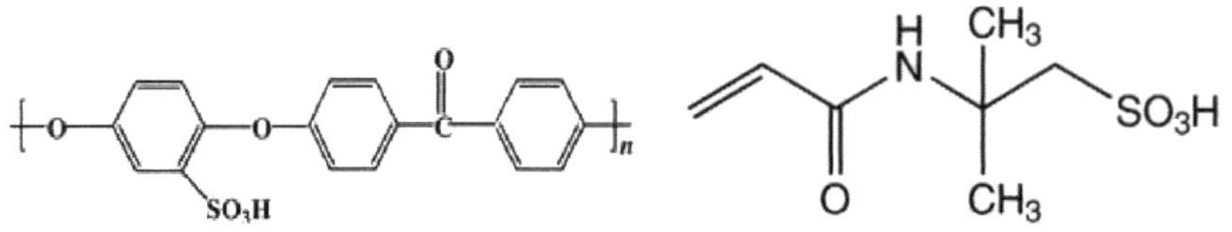

Figure 1 Structure of SPEEK and PAMPS

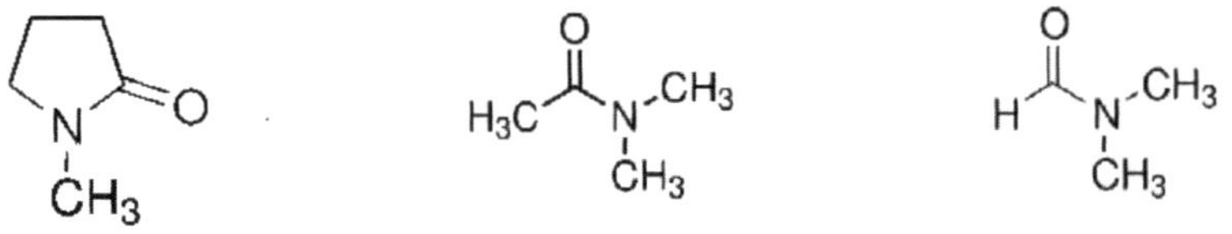

N-Methyl-2-pyrrolidone N,N-Dimethylacetamide N,N-Dimethylformamide

Figure 2 Structure of Solvents

Experimental Procedure

Membrane Preparation

SPEEK was first dissolved in solvents (DMAc and NMP) and stirred for 2hours at 70°C. PAMPS was then added and the blend was stirred for 24hours. The solution was sonicated for various periods of time ranging from 10 to 90minutes using an Ultrasonic bath. In untralsonic bath, the final temperature achieved for a given time of sonication depends on the tempearutre of the laboratory (summer or winter). It is therefoe genrally advisable that maximum temperature of the bath is determined and maintains (as the so - called equilibrium tempearture), when operating continuously under ambient conditions (Santos et al. 2009). During the sonication, equilibrium temperature was maitained by simply filling the bath with water that was preheated to sample temperature.

Membrane Characterization

Water uptake and morphological tests were performed. Water uptake was carried out by first vacuum drying all membrane samples at 100°C until constant weights were obtained. Dried samples were then immersed in deionized water for 24h at room temperature (25°C). Samples were taken out of the water after 24h, wipped to remove surface water and then weighed immediately. Water uptake was calculated using Eq. 1.

$$water - uptake = \frac{W_{wet} - W_{dry}}{W_{dry}} \qquad \text{Eq. 1}$$

Surface morphology of the membrane was done using scanning electron microscopy (SEM). Thermogravimetric analysis (TGA) was done from room temperature to 350°C using TA Q600 analyzer at heating rate of 20°C/min under nitrogen atmosphere with flow rate of 20mL/min using ~ 8 -10mg membranes samples.

Results and Discussion

Membrane Morphology by Scanning Electron Microspcopy (SEM)

Results of surface morphological analysis performed on membrane were shown in Figure 3 and 4. The overall pictures of this figure depicts a dense membrane with some surface defects observed for most of the sonicated samples. For example, Figure 3A showed some kind of layered arrangement with no visible pores at the surface. Also, the PAMPS seems to be well distributed throughout the SPEEK. The micrograph in Figure 3B can be observed to be rough and possesses some uneven ramdom distribution of PAMPS agglomerates indicating that the distribution of PAMPS have been disturbed by the 10minutes sonication. An increase in sonication time to 20 minutes was observed to improve the distribution of PAMPS. This was however accompanied by some cracks believed to be due to the slight increase in the viscousity of the blend solution. The vicousity increase was brought about by the excape of some fraction of the solvent during sonication. As the sonication time was increased, PAMPS was observed to become more evenly distributed and the cracks become more visible and larger in size (Figure 3D to F). SEM images of samples at sonication time of 10 and 30 minutes were reproduced in higher magnification of 10μm (Figure 4) in order to further revealed the effect of sonication on the morphology of the somposites. Similar morphological behaviour has been reported in literature, though for inorganic/organic clusters of nanoclay/epoxy composites membranes (Lam et al. 2005).

Water Uptake

Water uptake have been reported to be affected by the degree of sulfonation, membrane pretreatment, hydration, ambient relative humidity and water temperature (Xing et al. 2004; Wu et al. 2006). Figure 5 showed the effect of sonication time and solvent type on water uptake that was measured at room temperature (25°C). For both solvents, water uptake is same at zero sonication time. There was however a further decrease in the water uptake for NMP at 20 minutes sonication time before it beging to increase. Thus, the best sonication time for both solvent lied between 10 and 20 minutes with respect to water uptake. Interestingly, this result correspond to that reported by Lam (2005) on the effect of ultrasonication on microhardness of composites.

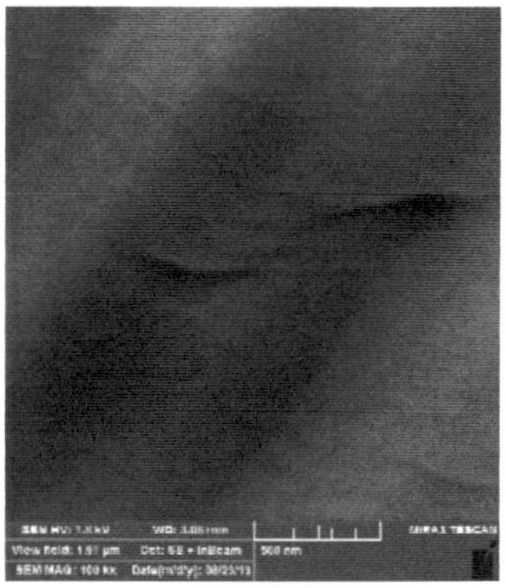

A) No sonication

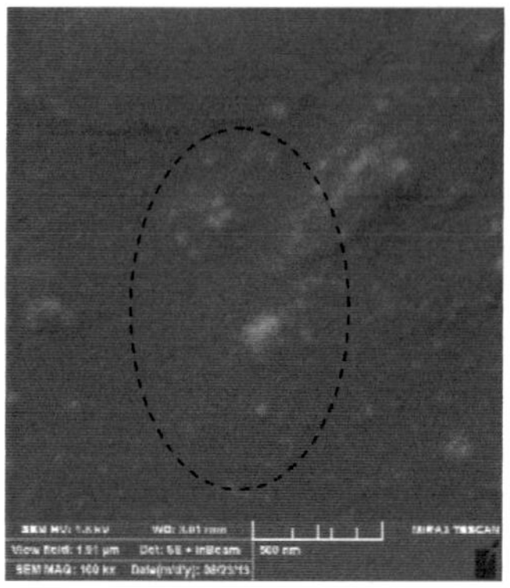

B) 10 minutes sonication

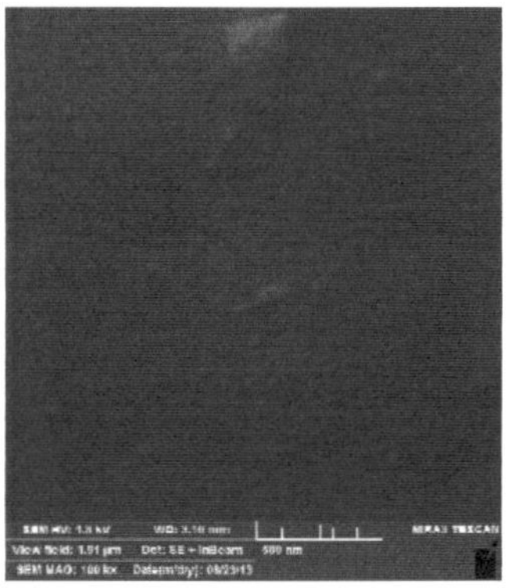

C) 20 minutes sonication

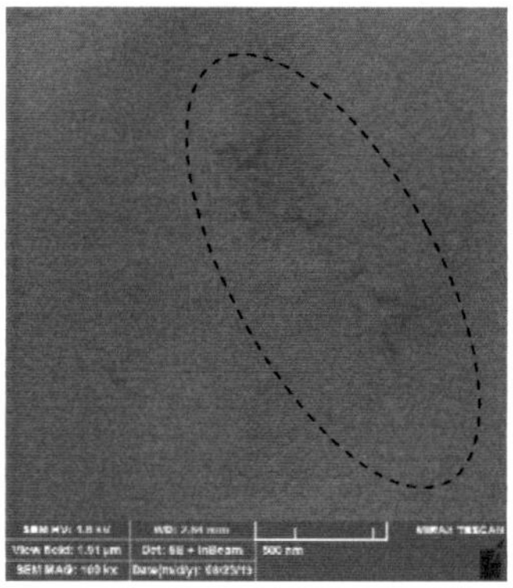

D) 30 minutes sonication

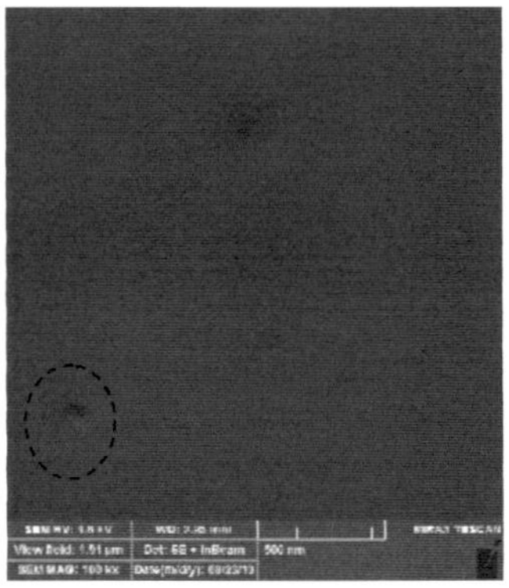

E) 60 minutes sonication

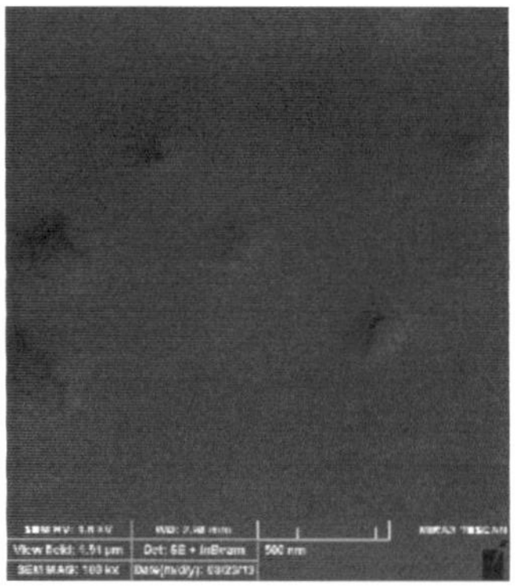

F) 90 minutes sonication

Figure 3 SEM micrograph of SPEEK-PAMS composite with various times of sonication with magnification of 500nm

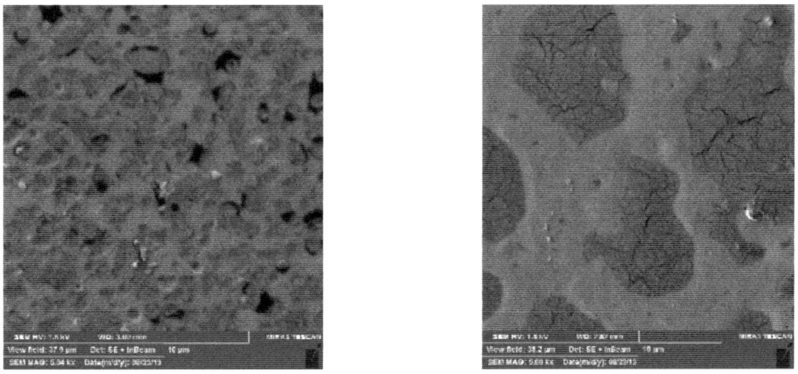

A) 10 minutes sonication B) 30 minutes sonication

Figure 4 SEM micrograph of SPEEK-PAMS composite with various times of sonication with magnification of 10μm

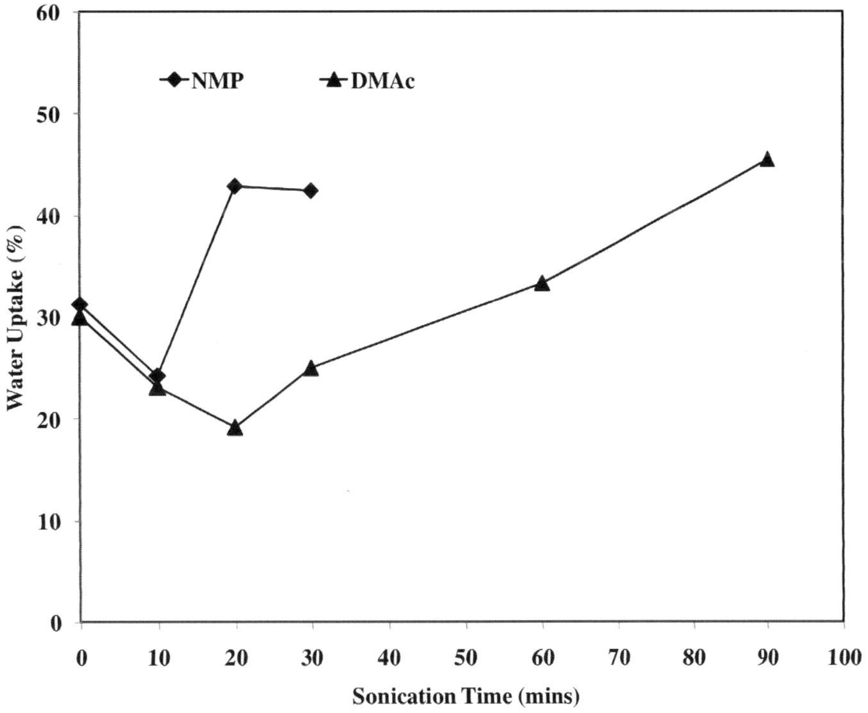

Figure 5 Water uptake of SPEEK-PAMS (3wt%) composite with variuos sonication times

Thermogravimetric analysis (TGA)

The difference in membrane performance resulting from the use different casting solvents have been attributed to the presence of residual solvents in the membrane samples after evaporation (Joly et al.

1999). TGA was thus used to quantify the residual solvent and evaluate its effect on the membrane performance as has been previously demonstrated in lieterature (Recio et al. 2007; Iqbal et al. 2008). Figure 6 shows the TGA curve for samples prepared from NMP and DMAc with different post vacuum heat treatment conditions. From this figure, weight loss can be attributed to evaporation of variety of substances, including solvents, from the samples (Rich et al. 1997). Table 2 contains the solvent and its boiling points, and weigth loss at the boiling point range that were extracted from TGA results. For sample heat treated for 24hrs, the overall percentage weight loss up to 204°C was about 8wt% of membrane sample for NMP while it is about 6.3wt% for DMAc. These will definitely include compnenents such as water, solvents and other unidentified substances. To quantify residual solvents, weight loss was estimated between 192 -214°C for SPEEK-PAMPS-NMP and between 154 and 176 for SPEEK-PAMPS-DMAc. The results obtained showed that the amount of the residual solvent in NMP is larger than that of DMAc by 1.25wt% (Table 1). To further investigate if this difference is responsible for higher value of water uptake for the NMP sample, membrane samples made from both solvents were further heat treated under vacuum for 72hrs. Residual solvent after this treatment was slightly higher in DMAc than NMP. Despite this, the value of water uptake of the NMP membrane is still higher than that of DMAc by approximately the same value as before the heat treatment (1.71wt%).

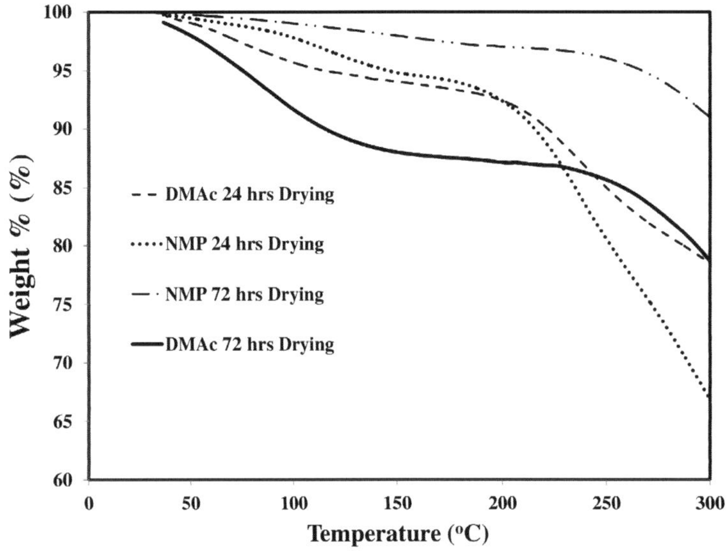

Figure 6 TGA of SPEEK-PAMPS-NMP and SPEEK-PAMPS-DMAc

Table 1 Residual solvent in SPEEK-PAMPS composites membrane samples

Membrane Sample	Solvent	B. Pt. (°C)*	Residual Solvent (wt%)	Water Uptake
DMAc 24 hrs Drying	DMAc	164.5 – 166	0.57	30.00
NMP 24 hrs Drying	NMP	202 – 204	2.85	31.25
DMAc 72 hrs Drying	DMAc	164.5 – 166	0.367	21.21
NMP 72 hrs Drying	NMP	202 – 204	0.217	22.92

*Boiling Points of Solvents

On a general note, results from Figire 1 – 5 showed that the best perfomance can be obtained for variety of sonication time depending the property of preference. Also, performance of DMAc solvent was observed to be better than that of NMP on the basis of water uptake. For this reason, DMAc was selected as the casting solvent for membrane preparation on which further studies will be carried out. Water uptake of DMAc membrane, prepared without sonication, was comapred with those reported in previuos literature. Table 2 illustrates the comparison between SPEEK-PAMPS and other relevant membranes previuosly reported in literature. In addition, this table serve to provide insight into the possible potentials of the composites. From this table, it was observed that the water uptake for SPEEK-PAMPS is within the range of values of other membranes on the table. Its water uptake is less than that of Nafion 112 which is a commercial membrane. The differences observed between the water uptake values of SPEEK – PAMPS and other membranes from the table can be explained based on three important factors - SO_3H group, the degree of sulfonation and experimental temperature (Gil et al. 2004; Shan et al. 2006; Wu et al. 2006). Water uptake of SPEEK – PAMPS is higher due to the presence of SO_3H group in the PAMPS molecules (Figure 2). SO_3H group have been reported to enhance membrane acidity and hydrophilic properties (Shan et al. 2006). Infact, the higher the SO_3H group the larger the water uptake. The degree of sulfonation has been reported to affect the water uptake of membranes. The higher the degree of sulfonation, the larger the value of water uptake (Gil et al. 2004). Thus, water uptake of SPEEK – PAMPS is lower than SPEEK -3 due their different degree of sulfonation, 53.3 and 72% respectively and the experimental temperature 24 and 60°C, respectively. The experimental temperature of SPEEK – 2 is close to that of the composite. However, the results cannot be compared in this case since the DS of SPEEK – 2 was not known. Above explanation is also true for the values of water uptake for SPEEK – 1.

Table 2 Comparison of water uptake values obtained with those from Literature (Xing et al. 2005)

Membrane	Water Uptake (wt%)	Experiment Temperature (°C)	**D S (%)	Reference
Nafion 112	22.85	N/A	-	(Xing et al. 2005)
SPEEK – 1*	16.00	30	45	(Ren et al. 2005)
SPEEK – 2*	44.90	25	N/A	(Zhong et al. 2008)
SPEEK – 3*	85.00	60	72	(Wu et al. 2006)
SPEEK– 4*	15.00	N/A	N/A	(Handayani and Dewi 2011)
SPEEK/PAMPS (3wt%)	21.21			This work

*The numbering have been included for explanation purposes.
**DS means Degree of Sulfonation

Conclusion and Recommendations

Blend of SPEEK and PAMPS was prepared by solvent casting method and its performance was evaluated with respect to its morphology and water uptake. Effect of sonication time (10 – 90minutes) and solvent types (DMAc and NMP) were also investigated. Performance of membrane prepared using DMAc was observed to be better than that of NMP. There was no correlation between the sonication time and membrane properties, but it was observed that the sonication deteriorate the membrane morphology. Membrane prepared without sonication shows better surface morphology, whereas the best water uptake was observed between 10 and 20 minutes. Thus, other properties need to be investigated in order decide the best sonication time. Therefore, further investigation such as thermal stability, methanol permeability, ion exchange capacity and gas (O_2 and H_2) permeability of the membrane are recommended for a complete evaluation of the membrane performance for direct methanol fuel cell (DMFC) application. All these investigations are in progress in our laboratory.

Acknowledgement

Authors would like to thank the National Science, Technology and Innovation Program (NSTIP) of KACST for funding this research under project number 10-ENE1374-04. Also, authors extend their appreciation to Center of Petroleum & Minearls, Center of Excellence in Renewable Energy, Mr. Dan Jerry D.Cortes of the Nanotechnology Laboratory, Chemical Engineering Department, King Fahd University of Petroeleum & Minerals, Saudi Arabis for their supports.

REFERENCE

Florjańczyk, Z., et al. (2003). "Proton conducting electrolytes based on poly(2-acrylamido-2-methyl-1-propanesulfonic acid)." Electrochimica Acta **48**(14–16): 2201-2206.

Gil, M., et al. (2004). "Direct synthesis of sulfonated aromatic poly(ether ether ketone) proton exchange membranes for fuel cell applications." Journal of Membrane Science **234**(1–2): 75-81.

Handayani, S. and E. L. Dewi (2011). "Influence of Silica/Sulfonated Polyether-Ether Ketone as Polymer Electrolyte Membrane for Hydrogen Fueled Proton Exchange Membrane Fuel Cells." Internat. J. of Sci. and Eng. **2**(2): 26-30.

Iqbal, M., et al. (2008). "Solvent effect on morphology and CO2/CH4 separation performance of asymmetric polycarbonate membranes." J. Membr. Sci. **318**(1–2): 167-175.

Joly, C., et al. (1999). "Residual solvent effect on the permeation properties of fluorinated polyimide films." Separation and Purification Technology **16**(1): 47-54.

Lam, C.-k., et al. (2005). "Effect of ultrasound sonication in nanoclay clusters of nanoclay/epoxy composites." Materials Letters **59**(11): 1369-1372.

Mohr, J. M., et al. (1991). "Gas transport properties of a series of poly(ether ketone) polymers." Polymer **32, 13**: 2387-2394

Recio, R., et al. (2007). "Gas separation of 6FDA–6FpDA membranes: Effect of the solvent on polymer surfaces and permselectivity." J. Membr. Sci. **293**(1–2): 22-28.

Ren, S., et al. (2005). "Surface modification of sulfonated poly(ether ether ketone) membranes using Nafion solution for direct methanol fuel cells." Journal of Membrane Science **247**(1–2): 59-63.

Rich, D. C., et al. (1997). "Effect of cure conditions on Probimide 32 polyamide-imide." Journal of Applied Polymer Science **63**(9): 1113-1126.

Santos, H. M., et al. (2009). The Power of Ultrasound. Ultrasound in Chemistry: Analytical Applications. J.-L. Capelo-Martinez. Weinheim, Wiley-VCH.

Shan, J., et al. (2006). "Sulfonated polyether ether ketone (PEEK-WC)/phosphotungstic acid composite: Preparation and characterization of the fuel cell membranes." Pure Appl. Chem **78**(9): 1781 - 1791.

Swier, S., et al. (2005). "Sulfonated poly(ether ketone ketone) ionomers as proton exchange membranes." Polymer Engineering & Science **45**(8): 1081-1091.

Wu, H.-L., et al. (2006). "Preparation and characterization of poly(ether sulfone)/sulfonated poly(ether ether ketone) blend membranes." European Polymer Journal **42**(7): 1688-1695.

Xing, D. M., et al. (2005). "Characterization of Sulfonated Poly(Ether Ether Ketone)/Polytetrafluoroethylene Composite Membranes for Fuel Cell Applications." Fuel Cells **5**(3): 406-411.

Xing, P., et al. (2004). "Synthesis and characterization of sulfonated poly(ether ether ketone) for proton exchange membranes." Journal of Membrane Science **229**(1–2): 95-106.

Ye, Y.-S., et al. (2012). "Water Soluble Polymers as Proton Exchange Membranes for Fuel Cells." Polymers **4**(2): 913-963.

Zhong, S., et al. (2008). "Modification of sulfonated poly(ether ether ketone) proton exchange membrane for reducing methanol crossover." Journal of Power Sources **180**(1): 23-28.

Fixed Plant Facility Siting – A Mega Project Lifecycle Framework

M.F. Haider
mech_haider@yahoo.com
mhaider@suncor.com

Prepared for Presentation at
American Institute of Chemical Engineers
2014 Spring Meeting
10th Global Congress on Process Safety
New Orleans, LA
March 30 – April 2, 2014

Fixed Plant Facility Siting – A Mega Project Lifecycle Framework

M.F. Haider
mhaider@suncor.com
mech_haider@yahoo.com

Keywords: Facility Siting, Project Management, Mega Project, Consequence Analysis (CA), Quantitative Risk Assessment (QRA), Adaptive, Change driven, Agile methodology

Abstract

Facility Siting based on tools such as Consequence Analysis (CA) and Quantitative Risk Assessment (QRA) is often times successfully applied for Brownfield operating unit projects, projects with relatively smaller footprints and for projects lasting relatively short time-periods. Challenge remains to apply and realize the full potential and value of Facility Siting for Mega-Projects. This is mostly due to project complexity and length – this paper will introduce a Management Framework to successfully apply CA and QRA as part of Facility Siting for Mega-Projects. The paper will also outline key strategies to ensure successful (i.e. value adding) implementation of facility siting, suggest a list of activities, associated sequence and pitfalls to avoid. Project Management team (e.g. Project Sponsor, Project Engineers, Managers) of both owner/operator and contractor/vendor may find this paper more useful as it provides a broad overview (i.e. when to initiate, how to plan, execute and close such assignment to maximize value) including real life examples of facility siting process over the project lifecycle.

1. Introduction

The driving force for process safety has been primarily based on catastrophic events [1]. With an increasing number of incidents, the process industry and governments started to take measures such as Facility Siting initiatives with hope to minimize loss of life and property. In the 1970s and 1980s, some of the world's most tragic industrial accidents took place. The Flixborough explosion in 1974 was reinforced by that of the Seveso tragedy in 1976 [2]. The Macondo blowout in the Gulf of Mexico on April 20, 2010, killed 11 employees and led to an uncontrolled oil spill lasting 87 days. This blowout was the most significant offshore incident in the US to-date. Finally, the Fukushima Daiichi nuclear plant incident in Japan in March 2011 drew the attention of the global process and power industries and encouraged them to incorporate natural disaster risks in a hazard analysis study [3]. As part of inherently safer design, many companies now-a-days apply CA and QRA as part of facility siting for their new projects. Due to the scarcity of information at the beginning, the studies in many cases start later in the project lifecycle for them to be effective and value adding. This paper will focus more on how to manage such initiatives for new Mega-Projects over the technical intricacies of the CA/QRA processes. The intention of the paper is to enhance the Body of Knowledge publicly available for management team in as much non technical languaga as possible to make their greenfiled Mega-Project initatives a success i.e. safer for people, environment and property.

So, what is different for industrial Mega-Projects?

- Industrial Megaprojects, as the name implies, are very large. To provide a simple and simply applied definition, we are defining a megaproject as any project with a total capital cost of more than $1 billion (U.S. dollars) as measured on January 1, 2003 [4].
- Most often engineering and construction activities run in parallel and not in sequence
- Multiple sub-projects under the 'Mega-Project umbrella' run in parallel
- Complex – design is almost 'never' complete in early stages – goes through many iterations
- Progressive elaboration is evident
- Scarcity of critical workforce with mega-project experience
- Scarcity of information – requires project team to make assumptions for CA/QRA
- Timeline: 2-5 years

2. Key Strategies

2.1 Facility Siting Charter

It is not only important or good to do but critical that the stakeholders need to identify, understand and agree on the purpose and value to conduct facility siting for their projects. The Author's personal experience over the years has been in order to maximize value of Facility Siting the project team has to start 'early' to develop project Charter and start with the end in mind. As can be seen from the following generalized Figure 1 that as the company starts to appraise opportunities, it is

Figure 1: Stages and Milestones to apply CA/QRA in a Mega-Project

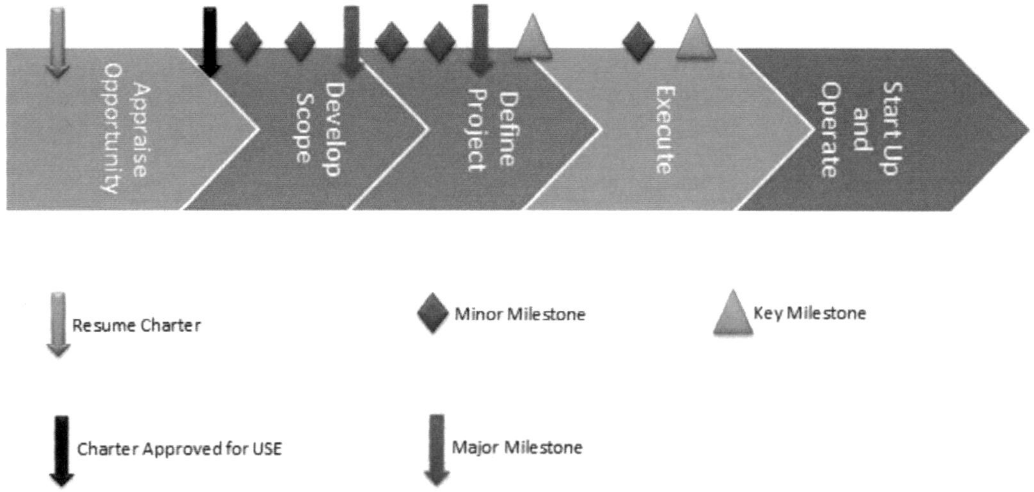

prudent to resume developing the Facility Siting Charter by bringing in the stakeholders together. Figure 1 shows a generalized project lifecycle for a typical Petro-Chemical Mega-Project. The

Figure also shows that the Key Deliverable is developed through a successive series of repeated cycles over the project life-cycle. The Milestones will be defined later in the paper. Part of starting with the end in mind is to clearly outlining ultimate goal/s, e.g. vision is to design a safer facility and utilize Mitigated QRA Report to formally document and communicate that facility is within the Risk Threshold or justification for accepting risk profile as is by Senior Management. The Vision in the Charter also provides opportunity to strike a balance between business goals and climbing the Risk Mountain – at each stage of increasing safety (decreasing risk). As part of striking a balance, the Project Engineer has to Identify options up-front if and/or when it becomes apparent that aggregate Residual Risk will not be within the Risk Threshold or acceptable so that team can put more effort in developing justification for either accepting or rejecting the risk profile by Management

Figure 2: Strike a balance between Business goal and Climbing the *expensive* Risk Mountain

2.2 Management Strategy

In order to successfully manage Facility Siting, CA and QRA for a Mega-Project, the stakeholders have to align on a management strategy. There are two primary management approaches for the purpose of managing such work:

1. Predictive or Plan driven
2. Adaptive or Change driven

The Author's past experience has been if the organization undertook very 'similar' projects in the near past (<5 years), critical experienced workforce is retained, Lessons Learned is readily available, and most importantly, the past project is still considered to be a success, it is advisable to use Plan driven approach. On the other hand, if the organization is introducing new technology, has limited past experience with similar project, involves relatively high number of stakeholders, capital cost > $1B, project life spans >2 years, it is recommended to utilize Change driven approach to maximize value. It is to be noted that it is not essential to meet all the suggested criteria mentioned above to pick one management strategy over the other.

2.2.1 Benefits of Adaptive or Change Driven Approach

In Adaptive/Change Driven Planning, project team intentionally repeats one or more project activities as the project team's understanding of the product increases ....Also known as ...agile methods .....intended to respond to high levels of change and ongoing stakeholder involvement [5]. The heart of Facility Siting involves making timely but informed decision. Some of the characteristics of this framework:

1. To add small incremental improvements that will deliver value to stakeholders
2. Milestones develop the product (e.g. Mitigated QRA Report) through a series of repeated cycles, while increments successively add to the functionality of the product
3. For each Milestone, a deliverable (read 'model') or set of deliverables will be completed. Future iterations may enhance those deliverables or create new ones. Each iteration incrementally builds the deliverables until the exit criteria for the phase are met, allowing the project team to incorporate feedback.

The biggest challenge with Mega-Project is scarcity of key information needed as inputs to develop scenarios and models. Some of the critical benefits, a project may gain by applying Change Driven approach:

1. The approach acts as Leading Indicator on Process Safety for Plant Layout - allows for multiple opportunities to make critical decisions on plant layout over the project life-cycle capitalizing on what is known at a given time. This lets the project team to use "what they know", to ensure "what they don't know" falls into the right place
2. Opportunity to make decisions over the project lifecycle not only works to save life but also helps with striking a balance between business needs and climbing the risk mountain
3. Aides design-development with progressive elaboration while producing final CA/QRA deliverable for the project

2.3 Execution Strategy

The main idea of Change Driven process starts with decomposing the known preliminary layout and process information into a set of requirements, work to be performed and their priority sequence, sometimes referred to as a *Product Backlog*.

It is advisable that the owner/operator forms a multi-use agreement (contract) with the chosen vendor. This way the plant areas under the Mega-Project umbrella can get their specific work done while the CA/QRA Project Team provides overall oversight and guidance.

At the beginning of an iteration, the team will work to determine how many of the highest priority items on the backlog list can be delivered within the next iteration. The time-length of each iteration should be less than 4 weeks. As shown in Figure 1, during "Develop Scope" and "Define Project", the length of each iteration could be even less than 2 weeks. This iterations (otherwise known as "Minor Milestone" as shown on Figure 1) will develop scenarios, assumptions and preliminary CA/QRA models. Assumptions are required due to lack of available quality information in the early stages of a project.

At the end of each iteration, the Minor Milestone/deliverable should be ready for review by the customers/stakeholders. This does not mean that the customer is required to accept delivery, just that the product should not include unfinished and/or incomplete features.

The sponsor and customer representatives should be continuously engaged with the other project stakeholders to provide feedback to Contractor on deliverables as they are created and to ensure that the *Product Backlog* reflects Sponsor/stakeholders' current needs. Below, you will find two examples using two different management strategies. For confidentiality purpose, the specific name/location is kept generic and few items were modified to suit.

2.3.1 Project X Facility Siting Using Change Driven Approach

A CA/QRA team was given the assignment to support design development during "Develop Scope" phase (Figure 1). The Sponsors already selected the site for the project. But the fixed plants were yet to be located on the site. The Vision for the initiative was for the QRA result to come under the organization Risk Threshold. Multiple EPC's were given to engineer different plant areas. The CA/QRA team selected a vendor to support them with defining the *Product Backlog* based on the preliminary Process Design Basis, intended product the facility is expected to sell, the feedstock necessary etc. The CA/QRA team had prioritized the Product Backlog based on hazard level of feedstock, preliminary PHA schedule and project schedule. This also involved obtaining feedback from different Process Engineering and Operations teams to collect necessary inputs to put together the assumptions and scenarios to be modeled within the 1st iteration.

Out of 25+ scenarios, only 7 perceived to be the most hazardous were modeled for CA that gave ample input (e.g. explosion, toxic and fire contours) for the CA/QRA team to work with the stakeholders to recommend location for specific process units on the given site. Note that, CA/QRA team integrated the potential consequence to understand local incident and global consequence events. At this time QRA was not performed. This was used as input for minor modification of preliminary plant layout. It is to be noted that the site plan was still preliminary after the first iteration but assisted different EPC accountable for their specific area to recognize where to locate the specific plants i.e. finalized key plans for the overall site plan.

With further design-development, more refined information became available from each EPC. This new information led to 130+ scenarios and was used for further CA/QRA iterations that created either new model and/or embellished the past ones. Each EPC was accountable for facility siting within their boundary while the CA/QRA team provided technical oversight and integration support for the overall project. The Sponsor/stakeholders were engaged on a regular basis to ensure that the Product Backlog reflects future modeling requirements to properly site facilities e.g. Fire Water Pump house, Tank Farms, site Medical Centre etc. Early engagement of CA/QRA team and higher management willingness to incorporate their feedback with the progressive elaboration of the project led to QRA result meeting the organization risk threshold and ultimately safer facility. Changes suggested based on CA/QRA work in earlier stage (i.e. Develop Scope, Define Project) were easier to incorporate into the design.

2.3.2 Project Y Facility Siting Using Plan Driven Approach

The sponsors for project Y selected the site for the project and decided to hold on to not perform CA/QRA work with the progressive elaboration of the project activities. This was mostly decided based on the fact that not all information was available to properly conduct the 'studies'. Fundamentally, the sponsors looked CA/QRA as deliverables requiring a more 'complete' plan. Once the Develop Scope (as referred in Figure 1) is almost complete, the sponsors decided to engage CA/QRA team and a vendor. By this time, the project had significantly high number of stakeholders. Many of the stakeholders were not very familiar with CA/QRA and the value it could bring to them. There was not much time to schedule and train the stakeholders as well to bring them up to speed. As a result, the stakeholders were unable to align on a Vision. The Vendor was asked to provide explosion, toxic and fire contour maps for specific scenarios and locations that were almost 'fixed' on the site plan at that point in time. The resultant contour maps showed significant consequence on nearby facility with over 700+ people. It was also noted that the plant area consuming the most hazardous substance would impact the nearby facilities significantly from consequence perspective. It should be noted that the result was only based on a selected few scenarios. By the time the project reached the "Execution" phase, more scenarios were added as part of the scope. But it was already too late to incorporate most of the recommended changes resulting from CA/QRA work. The project went through several area by area risk reduction workshops only to accept (by the company executives per organization risk matrix) the aggregate residual risk level which was also found to be not within the organization risk threshold.

2.3.3 Importance of Starting Early

Figure 3: Impact of variable based on project time [5]

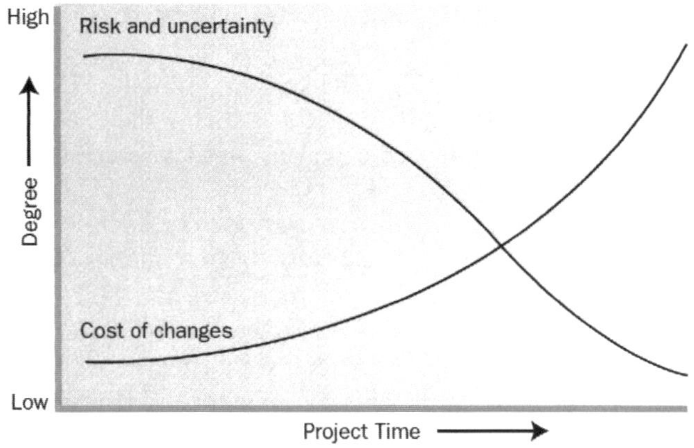

The above two examples highlight the importance of starting CA/QRA work early enough to realize and maximize the value CA/QRA brings to a Mega-Project. The high-level view provides common frame of reference for comparing projects for CA/QRA initiative, even if the projects were dissimilar. Change driven methodology allows for applying the known information at early stage in a way that supports laying out plant facilities to reduce risk and thus maximize potential value. While these characteristics remain present to some extent in almost all project life cycles, they are not always present to the same degree [5]. Change Driven framework, in particular, keeps stakeholder influences higher and the costs of changes lower especially in the early stages of the life cycle than in predictive/plan life cycles. The Figure 3 highlights a well-known fact of project management that even though risk and uncertainty remains high in the early stages of a project, cost of incorporating changes remains much lower. This translates into a key opportunity for project stakeholders to align on laying out the key areas on a site plan intelligently to reduce risk and maximize value by applying tools such as CA and QRA.

2.4 Key Decision Points

Facility Siting, from a management point of view, is all about bringing the stakeholders closer and find the 'right' spot for the facilities. Finding the 'right' spots requires making timely decisions. The challenge has always been to make timely decisions as the stakeholders more often find themselves forced to make those decisions based on information that are frequently of poor quality and at times uncertain, especially in the early stages of a project. At times, due to the scarcity of information, stakeholders rely on their past experience to make those decisions on plant layout. As a result, depending on the individual experience, two different teams for a project area may come up with two very different plant layout. CA/QRA work, if started early enough, will also depend on input that most often requires embellishment through assumptions to create needed scenarios and models. Even with this limitation, CA/QRA provides a key 'input' in allocating real estate for key facilities on a plant site at early stage of a project lifecycle. With each iteration and deliverable 'Change driven' methodology allows for multiple opportunities to make critical decisions on plant layout over the project life-cycle. It allows the stakeholders to decide when to 'sharpen the pencil' (e.g. PHAST to Computational Fluid Dynamics model), relocate facility or accept the consequence and/or risk, 'pull the plug' on further modeling/refinement and start developing justification, if required, to recommend management to accept/reject the aggregate risk profile.

2.5 Alignment with Process Hazard Analysis (PHA) Planning

CA/QRA results are a useful input to PHA (e.g. HAZOP, LOPA) activities. This is particularly true as projects move forward after establishing the plant layout on a high level and starts to conduct "Define Project" (Figure 1). PHA activities may also result in recommendations (i.e. new scenarios) that goes into the *Product Backlog* and requires project team to develop new models or refine old ones. Mega-projects have multiple sub-projects running in parallel most of the time, having an Integrated Schedule for all the reviews is a good strategy so that the right stakeholders are booked for the sessions. The integrated PHA schedule also supports prioritization of *Product Backlog* for CA/QRA activities over the lifecycle of the project.

2.6 Integration Strategy

It is critical for a Mega-Project involving multiple plant areas/silos to understand the BIG picture so that site risk can be dealt with on an aggregate basis. There is strong possibility of one area to impact other plants in the vicinity. It is recommended to review integration aspect and aggregated results on a regular basis. The following Figure 4 shows that the origin of an incident may reside in one particular plant area, while it may impact other areas in the vicinity.

Figure 4: Consequence of an incident in one area impacting multiple areas

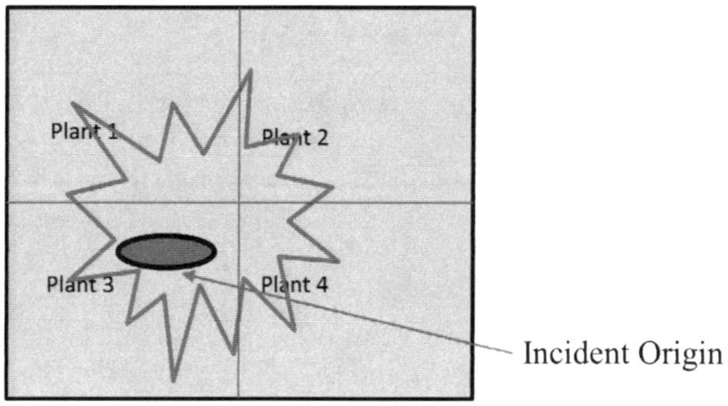

Recommended Strategy:

- Setup Strong Center - Task Force Single Point of Contact (SPOC)
- Collaborate – SPOC will promote Collaboration among Plants/Areas

3. List of Activities (suggested)

3.1 Overall Sequence

The CA/QRA work typically should start with the approved Charter as shown on Figure 1. A project Task Force i.e. SPOC is formed comprising of Project Engineer, Operations rep, Process rep and CA/QRA Technical Expert. The SPOC is ultimately accountable to produce the CA/QRA deliverable studies but also acts to support the design development and site plan finalization efforts in the early stages of the project.

The team collects 'input' such as Preliminary site Layout and design information, works with area process and operations team to identify scenarios, develop assumptions for known-unknowns and unknown-unknowns, develop modeling 'Product backlog', prioritize backlog based on PHA schedule and produce modeling output (i.e. deliverable) for stakeholder use. The stakeholder usually has the option of accepting or rejecting consequence. If the result of an iteration is acceptable within area causing the incident – communicate to outside stakeholders to ensure there is no cross boundary impact. If the result of the iteration is NOT acceptable to outside of the area – use the result as input to PHA for the areas to mitigate. Collaboration recommended between origin area and impacted area/s to identify appropriate mitigations. There is also a possibility that the consequence resulting from the iteration is not acceptable to the area causing the incident, in that case the area will initiate a PHA to mitigate the consequence. As part of developing mitigations, consider the following:

- Develop QRA if appropriate
- Perform Building Assessments for Critical Buildings
- Apply Cost-Benefit analysis to optimize risks/opportunities

- Determine Safe Haven locations and their characteristics

If Mitigation is not possible from Business perspective – perform As Low As Reasonably Practicable (ALARP) justification for Senior Management to formally accept the posed risk

- To apply mitigations on a site wide basis, review Milestone deliverables by taking on a BIG picture perspective
- Future iterations may enhance those deliverables or create new ones as new information is available. Each iteration incrementally builds the deliverables to produce the Mitigated CA/QRA report to support Facility Siting for the stage

The overall effort is expected to produce a maximum of two single reports for the overall site:

1. Mitigated version of QRA Report
2. and Consolidated CA Report

It is recommended to develop as-built, pre-start up CA/QRA reports prior to start-up. Table 1 and 2 outlines typical activities performed by SPOC and plant areas respectively once key process areas are established/finalized on Site Plan. Please note that the activities are project dependent and may warrant modifications to suit specific situation.

Table 1. SPOC Sequence of Activities to conduct Facility Siting

Initiative	Key Steps
Setup Facility Siting Team	Sponsor to support to setup a team (i.e. Project Engineer, Technical Expert, Operation, Process representative) to act as SPOC
Integration	Provide functional (technical and management) guidance to Areas
	Assess cross boundary impact/opportunity
	Communicate incremental outputs after each Area PHA/Risk Reduction
IFA versions (prior to "Define Project" stage)	Consolidated CA Report
	Mitigated and Integrated QRA Report
IFD CA/QRA (prior to "Execution" stage)	Re-validate IFA CA/QRA post PHA and vendor PHA
As-built versions	Post-oil version

3.2 Typical Area Activities

Table 2. Area (Typical) Sequence of Activities to conduct Facility Siting

Initiative	Key Steps
Plant 1 (Typical)	Perform Modeling and Calculations
	Conduct Risk Reduction Workshop (if appropriate) – Apply mitigations
	Incremental output for use based on reduced risk level
	Assess Buildings/facilities
	Identify Safe Haven/shelter Locations (if required)
	Estimate Area Project Scope impact
	Perform area scope updates/changes e.g. passive fire protection requirement
	Carry out QRA (if appropriate)
	Update Product Backlog
	Develop/update contract to support area specific activities

4. Pitfalls to Avoid

The following observations were noted by the Author during his time to manage Facility Siting scope:

1. Facility Siting does not replace good and common sense engineering practice e.g. locating a plant that uses high volume of hazardous material will significantly increase risks to other plants/areas (e.g. camp) within the vicinity
2. Evaluate effectiveness of 'Cost Reduction' philosophies as opposed to added risk e.g. different modularization techniques may decrease the plant footprint by reducing pipe lengths but makes the plant more compact and likely to increase severity of explosion consequence
3. Avoid developing models with hope to realize risk reduction
4. Sharpening the pencil for models most likely may not reduce risk
5. Avoid jumping into CFD or other similar methods without evaluating cost-benefit for undertaking such effort
6. Utilization of QRA results is much more controversial than the methodology. Watson [6] has suggested that QRA should be considered as an argument, rather than a declaration of truth. In this view – focus on accuracy over precision
7. CA/QRA Task Force should be part of Management of Change (MOC) process
8. Evaluate carefully before locating facilities e.g. tank farms close to main entrance/egress as part of Facility Siting

9. Keep the focus on VISION for the overall effort to avoid scope creep
10. Note the difference between Facility Siting as a Deliverable and tool for design-development activities

5. Conclusion

This paper presented Adaptive/Change driven framework to manage initiatives like facility siting/CA/QRA as part of Mega-Projects. Change driven framework does not mean "lack of planning or no planning". As demonstrated in the paper, this methodology starts with a vision and then develops the ultimate product both iteratively and incrementally when it is possible to define small incremental improvements that will deliver value to stakeholders. This allows making key decisions on laying out the facilities in early stages of a project rather than waiting for design to develop to a significant extent. Often times, project team wait too long for design to mature enough before applying CA/QRA to validate consequence impact and/or risk level. Most often, especially for Mega Projects, this leads to delays in executing the CA/QRA work properly at the right time to influence site layout. This delay in execution leads to scenarios when scope changes cannot be incorporated anymore. This translates into value erosion as project was unable to reap the full benefit of Facility Siting. The management framework recommended in this paper i.e. Change driven/adaptive framework allows for the stakeholders to start early in the project lifecycle to apply CA/QRA in laying out the plant facilities. The key is to start EARLY but START with the END in mind. Alignment among stakeholder on a VISION up-front is also important as this is expected to save life while striking a balance between business needs and climbing the Risk Mountain. Vision also allows avoiding scope creep while applying Adaptive/Change driven approach. It is highly recommended that stakeholders focus more on accuracy over precision. In the early stages of project lifecycle, quality information is scarce. Change driven/Adaptive framework gives you as the project stakeholder the opportunity to use "what you know" intelligently to ensure "what you don't know" falls into the right place.

6. References

[1] M. Mannan, A. Chowdhury and O. Reyes-Valdes, "A portrait of process safety: From its start to present day," College Station, 2012.

[2] M. Mannan, Lees' Loss Prevention in the Process Industries, Volumes 1–3 (3rd Edition), , 2005, Elsevier, 2005.

[3] K. McAndrews, "Consequences of Macondo: A Summary of Recently Proposed and Enacted Changes to US Offshore Drilling Safety and Environmental Regulation," in *Society of Petroleum Engineers, Americas E&P Health, Safety, Security and Environmental Conference*, 2012.

[4] E. W. Merrow, Industrial Megaprojects Concepts, Strategies and Practices for Success, New Jersey.: John Wiley & Sons, Inc., 2011.

[5] A Guide to the Project Management Body of Knowledge (PMBOK® Guide) – Fifth Edition, Newtown Square: Project Management Institute, Inc., 2013.

[6] S. Watson, "The Meaning of Probability in Probabilistic Risk Assessment," *Reliability Engineering and System Safety,* vol. 45, pp. 261-269, 1994.

348480 Global Energy and Transportation and Mobil Oil

Abstract:

Biofuels, Computational Methods, Engineering Economics, Polymers, Thermodynamics. Polymers-plastic, Thermodynamics- oil, Engineering Economics – the money it takes to finish the method of Global Engineering Research, Computational Methods – numerical input of analytical data, Biofuels – the transport fluids that utilized in more than one method from industrial usage to everyday living which enters the atmosphere. Exxon has completed a $600 million dollar deal for production of algae biofuel, amongst Mobil Oil has performed major research Global Energy and the effects of oil. Reading a most recent article on Mobil website.

348748 Characterization of Iron Phthalocyanine As the Active Material for Lithium Batteries

Abstract:

Results on the determination of the lithium-ion diffusion coefficient in the iron phthalocyanine (FePc) active material as a function of intercalation level, using a modified electrochemical impedance spectroscopy technique, will be presented. Also, developed thermodynamic functions such as the Gibbs free energy, entropy, and enthalpy changes for the lithium insertion reaction with FePc, derived from cell open-circuit voltage measurements as a function of intercalation level, are presented. Also presented are estimates of the maximum amount of lithium that can be intercalated into the FePc based on constant volume and temperature conditions.

349086 Design of a Free-Fall Reactor for Fast Pyrolysis of Waste Plastics

Abstract:

Thermal pyrolysis has earned significant reputation in the recovery of valuable chemicals, especially petrochemical based fuels from various materials, including biomass and plastics. Of many available technologies, *fast* pyrolysis was preferred in his study owing to its minimal heat and mass transfer limitations and better control on process conditions. The focus of this study is towards efficient conversion of waste plastics into gaseous products consisting primarily of alkanes, alkenes and aromatics. Lab scale experiments were performed using a micro reactor to determine favorable conditions, including reaction temperature and particle size that would result in maximum yield of ethylene. Using these results, a simple industrial size radiatively heated free-fall reactor was designed for the purpose of pyrolyzing waste plastics. The goal is to design a *fast* pyrolysis reactor that improves upon inefficiencies in traditional reactors while resulting in high yield of ethylene.

349732 Biosorptive Dehydration of Ethanol/Water Azeotropes Using Compound Starch-Based Adsorbent

Abstract:

Abstract: Recently, the interest in the production of bioethanol has greatly increased as a result of the increasing fossil fuel prices and associated environmental issues. Bioethanol is made from either grains or biomass and the broth contains 6 to 12 percent ethanol. In addition, water removal represents a serious problem as ethanol form azeotrope with water at atmospheric pressure. Distillation to water-free ethanol consumes 50 to 80 percent of energy used in the fermentation ethanol manufacturing process. Various techniques have been developed to break the azeotropic point, such as salting-out method, azeotropic distillation, extractive distillation, reactive distillation, adsorptive distillation and pervaporation. Among the methods, the adsorptive option is the most attractive, especially the biosorption, which has distinct advantages over conventional methods as it is environmentally friendly, highly selective, easily available, easy to operate, cost effective and reusable in repeated cycles in the treatment of aqueous organics.

In this work, a specially formulated compound starch-based adsorbent (CSA), which consists of corn, sweet potato and foaming agent, was developed. As an exploratory study, this research was focused on the investigation of CSA for biosorptive dehydration of aqueous ethanol and optimization of conditions for its maximum adsorption under various conditions of vapor feed flow rate, bed temperature and bed height through an orthogonal design method. Furthermore, to investigate the adsorption mechanism, dynamic saturated adsorbance of CSA was studied and characterization of the adsorbent before and after adsorption, as well as after regeneration, was performed.

In the first step, the net retention time and separation factor of ethanol and water were obtained using inverse gas chromatography (IGC), the results indicated that the net retention time of water on CSA is significantly longer than that of ethanol which attested the feasibility of this adsorptive separation purpose and low temperature could be propitious to the process which was consistent with the traditional theory of adsorption.

Then, through a batch adsorption experiment, the optimum adsorption condition was determined under vapor feed flow rate, bed temperature and bed height by orthogonal design method, which indicated that the optimization condition for ethanol was at vapor feed flow rate of 1.1g/min, bed temperature of 83 and bed height of 25cm. The packing density was 445kg/m^3. The residence time is determined by vapor feed flow rate and bed height. The residence time is more influenced than bed temperature. The longer the residence time, the more the output is. But it cannot be ignored that the adsorption selectivity may decreases. Low bed temperature is benefited to the adsorption process which agrees with the result of IGC. Five cycles experiments of adsorption-regeneration were carried out at the optimum adsorption and regeneration condition, the results showed that there is practically no change in the regeneration efficiency of CSA after the cyclic process adsorption–regeneration. It is due to the properly chosen regeneration conditions assuring complete desorption of the water adsorbed without changes in the CSA structure. This unchanged adsorption capacity after a number of cycles demonstrates reusability of the adsorbent.

To study the adsorption mechanism, experiment was conducted to study the dynamic saturated adsorbance of CSA. The water adsorption capacity of CSA has been found to be similar to some molecular sieves, which is about 0.15g/g. Besides, field emission scanning electron microscopy (FESEM) and mercury porosimetry (MP) were applied to investigate the change of surface morphology and microstructure of CSA before and after the adsorption, as well as after regeneration, as shown in Fig.1 and Fig.2. The FESEM images showed that compared with the particles before the adsorption, they became plump and the void spaces became smaller after adsorption, which indicated that the CSA was saturated with water. However, the CSA particles after regeneration became wizened like the nature particles, which demonstrated that almost all the water had been desorbed from CSA. The pore size distribution (PSD) indicated that compared with the PSD curve before adsorption, the peak at 6000nm disappeared and the peak at 17000nm became remarkably smaller after adsorption, which manifested that CSA is mainly composed of macropores which play a major role in the adsorption. However, the peaks at 6000 and 17000nm reappeared after regeneration, and the curve looked exactly the same with that before adsorption, which further demonstrated that CSA after regeneration can be totally regenerated and recovered to the original structure.

These results indicated that the starch-based adsorbent can be considered as an effective and low-cost bio-sorbent to dehydrate alcohol.

349781 Purification of 2-Amino-1-Phenylethanol Enantiomers By a New Technique Combining Distillation and Crystallization

Abstract:

Stripping crystallization (SC) is a new separation technique which combines distillation and crystallization. Basically, SC is operated at a triple-point condition, in which the liquid is simultaneously vaporized and crystallized due to the three-phase equilibrium. In essence, the process is continued until the liquid phase is completely eliminated and only the pure solid crystals remain in the feed. Thus, crystal washing is not required since only pure solid crystals remain in the feed and no impurities are adhered on the crystal surfaces at the end of SC. In the previous research, SC has been successfully applied to separate p-xylene from the mixed xylenes by Shiau et al. (2005, 2008).

Phenylethanolamine, also known as 2-amino-1-phenylethanol (2-APE), is an important basic skeleton in many adrenergic drugs and is often used as ingredients of cold medicines and anorectics. As the biological activity and pharmacokinetics of the enantiomers of a chiral drug can be different, the objective of this research is to apply SC to purify R-2-APE from the chiral 2-APE mixture. The experiments show that no crystallization occurs during SC without seeding due to subcooling of the liquid enantiomers. On the other hand, seeding can eliminate subcooling during SC, leading to the R-2-APE crystals gradually formed preferentially on the seeds. Thus, seeding with magnetic stirring can be successfully employed to purify R-2-APE from the 2-APE enantiomers.

1. Shiau, L. D., Wen, C. C. and Lin, B. S. Separation and purification of p-xylene from the mixture of m-xylene and p-xylene by distillative freezing, *Ind. Eng. Chem. Res*, **44(7)**, 2258-2265, 2005.

2. Shiau, L. D., Wen, C. C. and Lin, B. S. Separation of p-xylene from the multicomponent xylene system by stripping crystallization, AIChE Journal, 54(1), 337-342, 2008

350841 Fabrication of Low Cost Insulating Material from Kaolin Clay for Construction Purposes

Abstract:

A new low cost, thermal insulator, construction mineral polymeric material was synthesized from local Saudi Arabia resources using Inorganic Mineral Polymerization technique (MiP). In this technique, the mineral polymer is produced from the following components: clay mineral rich in koalinite, sand, sodium hydroxide solution, and water. Inorganic polymerization experiments are conducted by mixing known amounts of clay minerals and sand in a variable speed stainless-steel mixer for some period of times until a homogenous solid mixture is obtained. Sodium hydroxide solution is added to the mixture and the polymerization reaction takes place at different mixing speeds and times. The produced mixture is divided into a number of samples of known weights. Each sample is moulded in a stainless-steel cylinder at a medium pressure. Then the moulded samples are cured in an oven for 24 hours. The curing temperature was 80°C. The mineral polymerization conditions such as the ratio of clay to sand filler, particle size of the clay, particle size of the sand, concentration and amount of sodium hydroxide, speed and time of mixing were optimized to obtain stable and hard mineral polymer. Compressive strength for different specimens of fabricated mineral polymers was tested under dry and wet conditions. Values of compressive strength with values ranging from 11.5 MPa to 27.9 MPa were obtained.

351141 Thermodynamic Study of Binary PAH (Anthracene + Phenanthrene) Solid Mixtures

Abstract:

Polycyclic aromatic hydrocarbons (PAHs) are common components of many materials, such as petroleum and various types of tars. They are generally present in mixtures, occurring both naturally and as byproducts of fuel processing operations. It is important to understand the thermodynamic properties of such mixtures in order to better understand and predict their behavior (i.e., fate and transport) in the environment and in industrial operations. The Knudsen effusion technique was combined with gas chromatography and differential scanning calorimetry to measure the vapor pressure, composition, and phase behavior of binary anthracene and phenanthrene mixtures at near ambient temperatures. Mixtures of anthracene and phenanthrene exhibited non-ideal mixture behavior: In mixtures with less than 80% phenanthrene, this more volatile component preferentially sublimed and the vapor pressures decreased as the mixtures became enriched in anthracene. The vapor pressure data do not follow Raoult's law (that is, ideal mixture) predictions. Mixtures with high phenanthrene content (> 80%) showed azeotrope like behavior, in that anthracene and phenanthrene sublimed at nearly constant vapor pressure and molecular ratio. These results confirm that the behavior of few-component PAH mixtures does not follow ideal mixture models. Hence, as tars and petroleum (and potentially other organic compound-containing mixtures) weather to few component mixtures over time, complicated, non-ideal phase behavior might be observed.

351160 How Confident Are You That Your Major Accident Risks Are Under Control?

Abstract:

Major accidents in the process industries including those at Texas City , Buncefield UK and in the Gulf of Mexico have highlighted the critical role played by senior managers and executives. Effective leadership is essential if companies are to develop a positive safety culture that remains constantly vigilant towards the risk of a major process safety incident. In the UK the HSE has made leadership a key priority for the high hazard industries and there is a cross-industry initiative to develop and deliver training focussed towards board members responsible for process safety performance.

This paper explores the leadership failings that contributed to recent major accidents, and the cost to both the company reputation and the individuals involved. Key principles are discussed that can help improve leadership in the major hazard industries. These include; attitudes and decisions of senior manager affecting the investment strategy and the safety culture of the organisation, reinforcing the importance of safety by personal example, thorough understanding of major accident hazards and key risk control systems, investigating process safety incident and near misses to find the underlying causes, developing world class safety management systems, and identifying weaknesses in these systems using targeted performance indicators.

The author bases this paper on the development of a training standard for 'Process Safety Leadership' involving a cross-industry 'expert panel'. Training of senior executives in compliance with this standard has been carried out from late 2011 and feedback from these courses and the key requirements for effective leadership will be discussed.

351236 Sil Determination: Shortcomings with the Use of LOPA

Abstract:

Proposed Topic Area: Managing Risks

For a number of years, industry has been trying to apply the principles outlined in IEC 61508. In the process sector, this has resulted in a focus on the requirements of IEC 61511 (2003) and ANSI/ISA 84.00.01 (2004). These are risk based standards. They involve the identification of potential hazardous events, the causes that could lead to each event and the necessary risk reduction for each event. Where the risk reduction involves an instrumented function (e.g. a trip or an interlock), the required performance of the instrumented function can described in terms of the average probability of failure on demand. In the standards, the performance for such instrumented functions is assigned to one of four Safety Integrity Levels. The assessment of what performance level is needed in respect to a specific hazardous event is referred to as SIL Determination. In the field of SIL Determination, IEC 61511 provides examples of a number of different approaches – these are shown in IEC 61511 Part 3. These include Event Trees, Risk Graphs, Safety Matrices, Fault Tree Analysis and Layer of Protection Analysis (LOPA).

In the first few years, following the publication of IEC 61508 Edition 1 in 2000, industry generally embraced the use of Risk Graphs as the method of choice. However, over the subsequent few years, many companies using this method found that it was (a) giving higher safety integrity level targets than really needed and (b) was less flexible than other methods and consequently more difficult to use in many real situations. This has resulted in a general move to other methods and in particular to the use of Layer of Protection Analysis. Indeed, LOPA may now be the predominant "Method of Choice" for SIL Determination. As a method, LOPA gives the impression of simplicity and appears quite straightforward. It almost invites anyone with a minimum of understanding of the method to be a practitioner.

However, there are a number of shortcomings in the use of the LOPA method that have become evident through its use across the globe.

This paper will review the application of LOPA and provide insights into what may be seen as the proper use of the method and its core principles. It will also highlight various shortcomings in use that have been found during expert reviews carried out of a number of existing studies. The paper will be of benefit both to those who are new to the LOPA method, but also of value to those who have been using the method for some time.

Shale Oil and Gas – Wellhead to Fuel

Dr. Lawrence Kremer - Baker Hughes
AIChE - Hilton New Orleans Riverside
April 1, 2014

14 AIChE
Spring Meeting, New Orleans, LA

BAKER HUGHES

Outline Unconventional Shale Oil

- Properties of shale oils
- Problems
  - Transportation
  - Refining

14 AIChE
Spring Meeting, New Orleans, LA

BAKER HUGHES

General Characteristics of Shale Oils

- API gravity typically varies 30-55+
- Very paraffinic
- Low sulfur
- Can have hydrogen sulfide and mercaptans
- Low TAN
- Very low asphaltenes
- Light crude with very little residual material
- Many samples show great quality variation

Factor	Bakken	WCS
API	39.7	21.6
S%	0.25	3.44
Light ends	2.6%	
LSR	3.5%	3%
Light Nap	12.4	3.0
Heavy Nap	9.7	5.0
Jet	10.5	10.0
LFO	27.6	12.0
Gasoil	27.2	37.0
VacTar	6.7	29.8

14 AIChE Spring Meeting, New Orleans, LA

BAKER HUGHES

Shale Oil Property Comparison

Parameter	Eagle Ford	Bakken
API	52	40.8
TAN (g KOH/g)	< 0.05	0.09
Sulfur (wt %)	<0.2	0.304
Asphaltene (wt %)	0.1	0.41
Resin (wt %)	1.6	4.95
Filterable Solids (PTB)	225	76

14 AIChE Spring Meeting, New Orleans, LA

BAKER HUGHES

Utica Shale Crude - Three Phase

14AIChE
Spring Meeting, New Orleans, LA

BAKER HUGHES

Eagle Ford Crude Samples

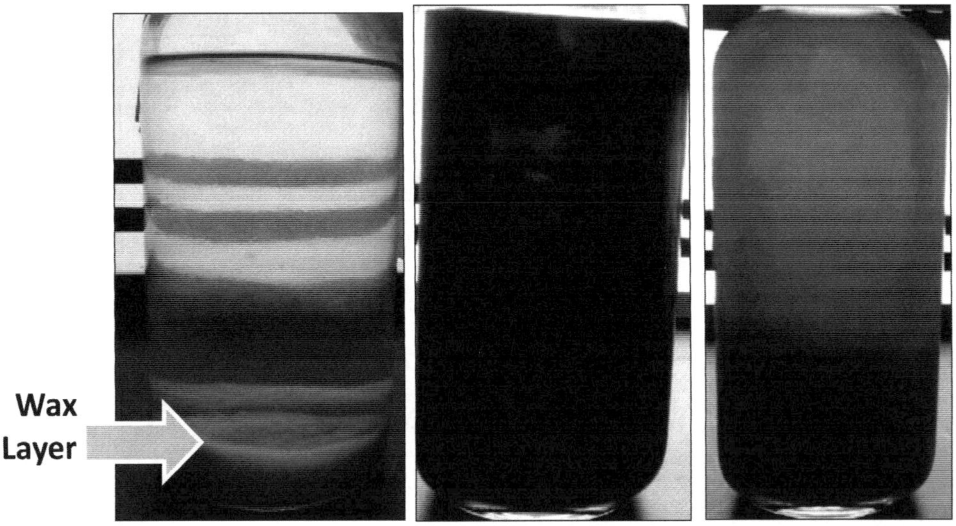

14AIChE
Spring Meeting, New Orleans, LA

BAKER HUGHES

Analytical Breakdown

Value	Yellow	Red	Black
API	55.0	44.6	52.3
TAN	<0.05	0.07	<0.05
Sulfur	<0.20%	<0.2%	<0.2%
Na	1.0 ppm	1.6 ppm	1.6 ppm
K	0.3 ppm	0.4 ppm	0.5 ppm
Mg	3.4 ppm	2.9 ppm	3.0 ppm
Ca	2.6 ppm	2.8 ppm	3.8 ppm
Asphaltene	0%	0%	0.1%
Resin	0.5%	3.2%	1.6%
Filterable solids	176 PTB	295 PTB	225 PTB
P	5.3 ppm	4.6 ppm	5.0 ppm

14AIChE Spring Meeting, New Orleans, LA

BAKER HUGHES

Micrograph of Sludge

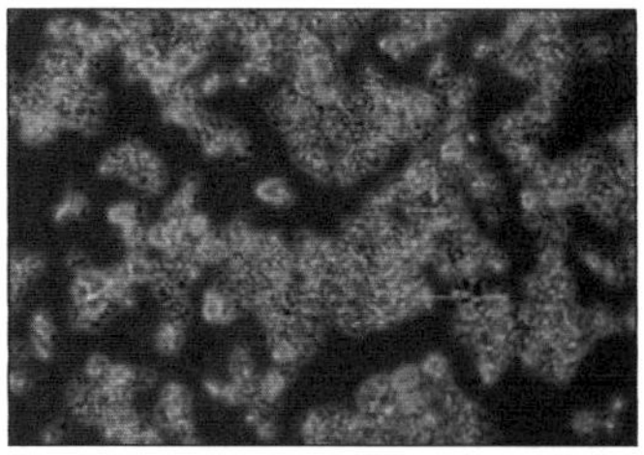

100 X

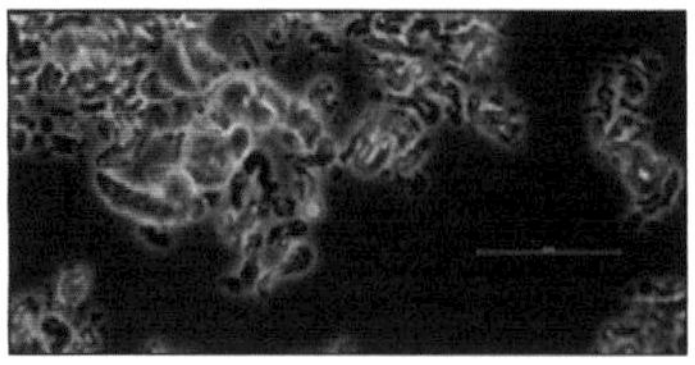

400 X

14AIChE Spring Meeting, New Orleans, LA

BAKER HUGHES

Paraffin Analysis of Sample

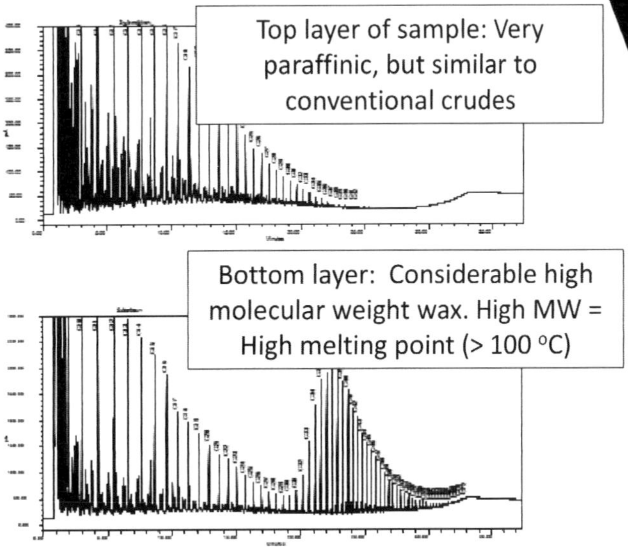

14 AIChE Spring Meeting, New Orleans, LA

BAKER HUGHES

Transport of Shale Production

- Infrastructure does not exist to meet needs
  - Most shale oil moved by rail or trucks
  - Limited pipelines expanding to meet transportation needs
- Unit trains to East Coast, West Coast, and Gulf Coast
- Problems:
  - Tankers and barges with heavy wax deposits ROB
  - Tight emulsions that look like oil or white condensate but high BS&W
  - Cars not uniformly treated for H_2S
  - Corrosion in rail cars

14 AIChE Spring Meeting, New Orleans, LA

BAKER HUGHES

Viscosity of ROB Before Treatment

ROB from untreated compartment

14AIChE Spring Meeting, New Orleans, LA

BAKER HUGHES

Eagle Ford Shale Oil Wax Dispersion Testing

Several products were effective in dispersing the paraffin wax in an Eagle Ford shale oil sample at ambient temperature for over a month.

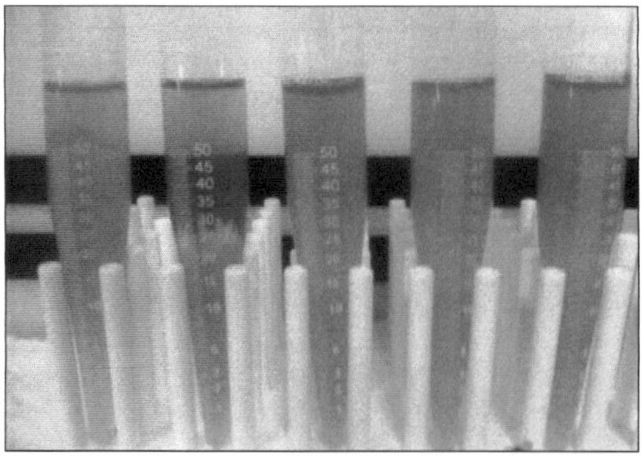

14AIChE Spring Meeting, New Orleans, LA

BAKER HUGHES

Product 'A' Effect on Wax Deposition (Utica)

Blank **200 ppm** **400 ppm**

14 AIChE Spring Meeting, New Orleans, LA

BAKER HUGHES

Refinery Issues

- Crude Blending
- Fouling
- Corrosion
- Product quality

14 AIChE Spring Meeting, New Orleans, LA

BAKER HUGHES

Crude Unit Fouling

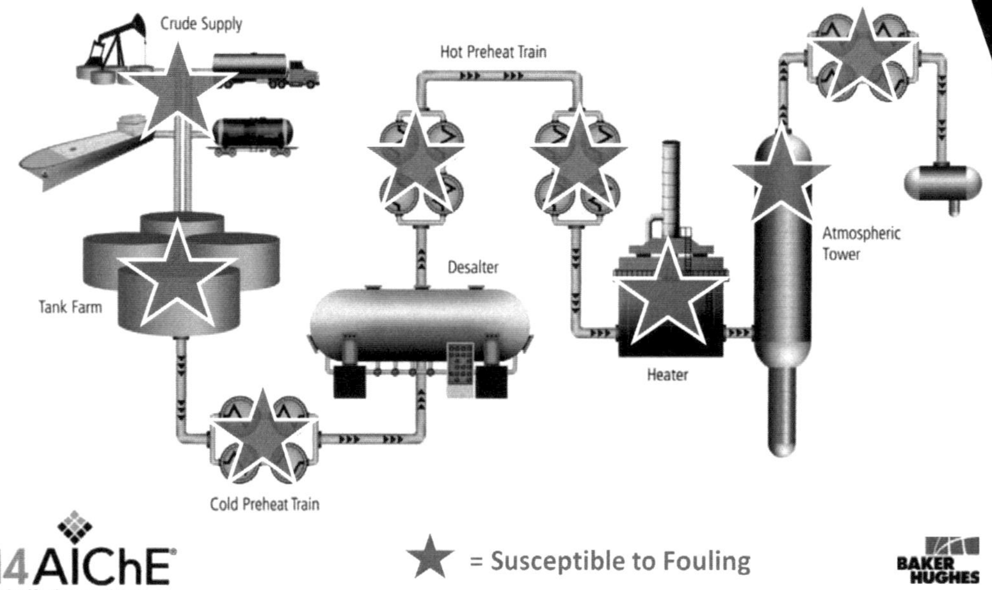

14AIChE Spring Meeting, New Orleans, LA

BAKER HUGHES

Asphaltene Precipitation

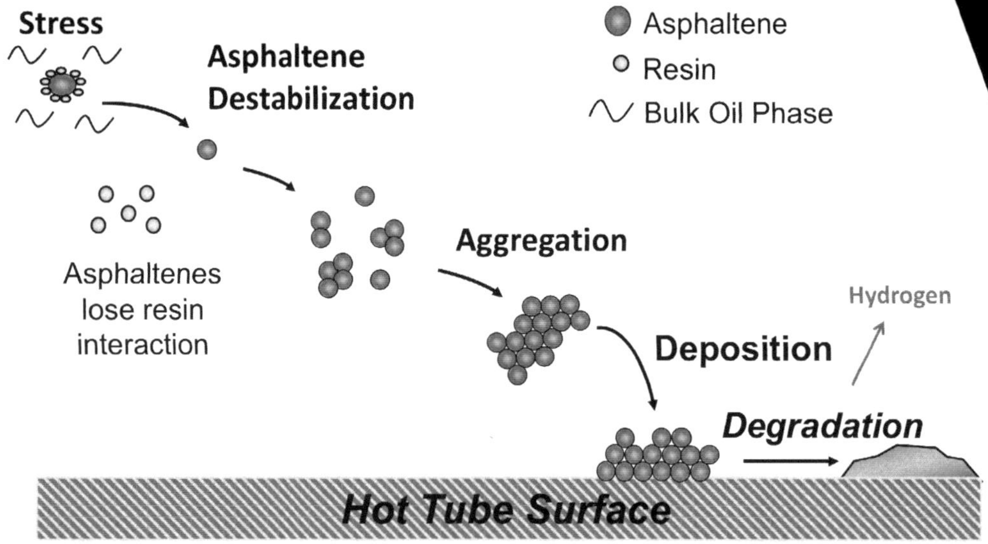

14AIChE Spring Meeting, New Orleans, LA

BAKER HUGHES

Asphaltene Stability Index Test

Field ASIT services™ technology measures the onset of the flocculation of the asphaltenes with high accuracy by inducing the asphaltene precipitation via titration with a paraffinic solvent.

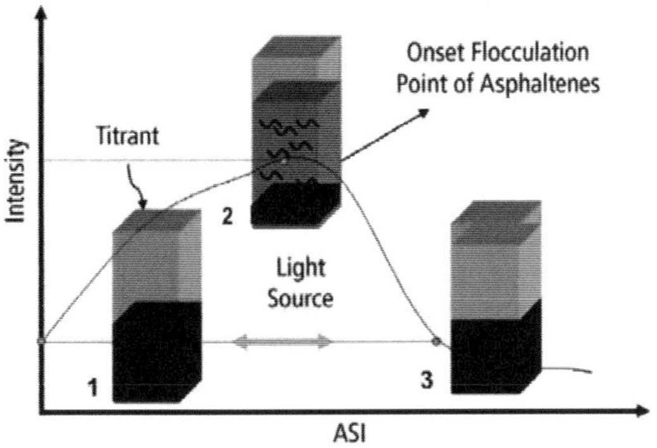

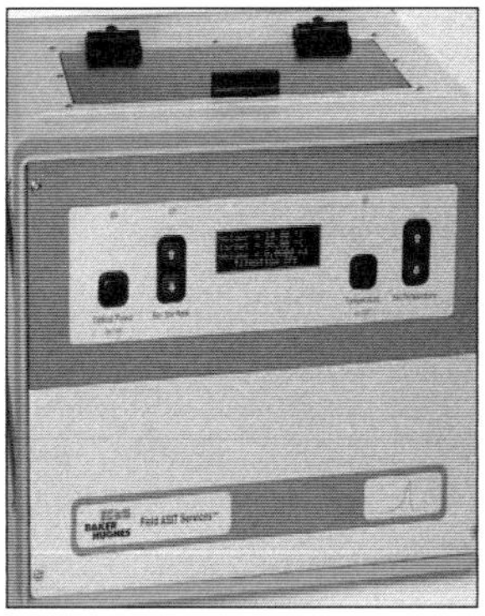

14AIChE
Spring Meeting, New Orleans, LA

Field ASIT services is a trademark of Baker Hughes Incorporated

BAKER HUGHES

ASI Shows Eagle Ford Can Destabilize Crude

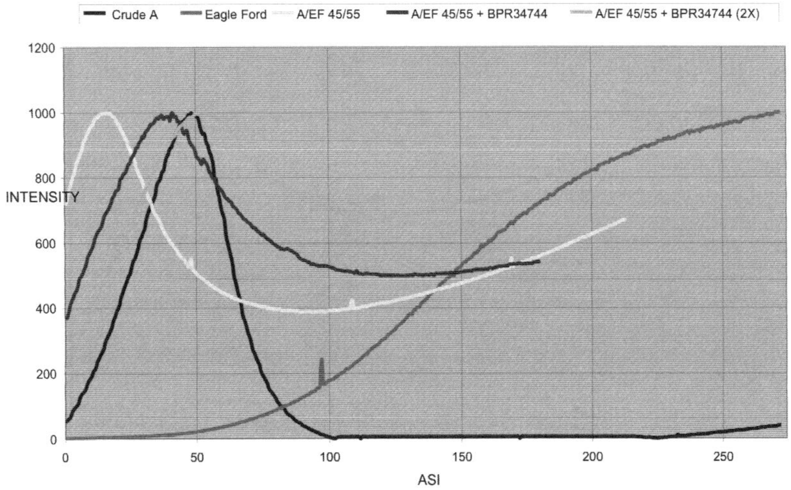

14AIChE
Spring Meeting, New Orleans, LA

BAKER HUGHES

Tower Tray Fouling - Phosphorus

- Seen by refineries running Bakken, Eagle Ford, and WTI
- Operationally manifested as increasing tower dP
  - Typically in the kero/jet zone
- Tower inspection shows deposits on trays in kero/jet zone
  - Some deposition may also show a few tray levels above and/or below kero/jet zone
  - Deposit scale typically tightly bonded to tray surface restricting flow and valve function

Above kero/jet zone

In kero/jet zone

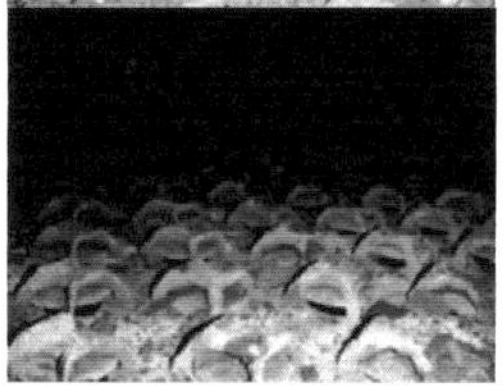

Below kero/jet zone

14AIChE Spring Meeting, New Orleans, LA

BAKER HUGHES

Shale Oil Impact – MEA Contamination

Corrosion and Fouling

- Shale oils transported by truck, rail, barge.
- Many shipments are treated with H_2S scavengers:

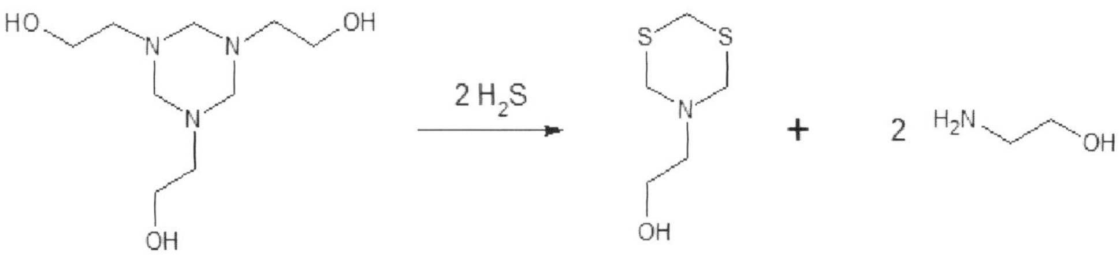

- Potential Impacts:
  - H_2S scavengers leave byproduct amine in crude oil (MEA, MA).
  - Tramp amines distill to OH system, create amine-HCl salts:
  - Corrosive to system metallurgy.

14AIChE Spring Meeting, New Orleans, LA

BAKER HUGHES

Impact of Tramp MEA on OH Water

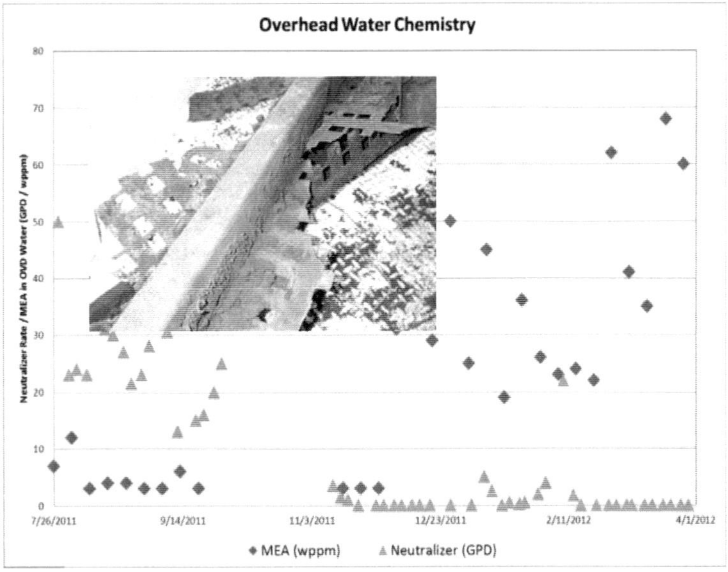

14 AIChE
Spring Meeting, New Orleans, LA

BAKER HUGHES

Salt Risk w/MEA Concentration

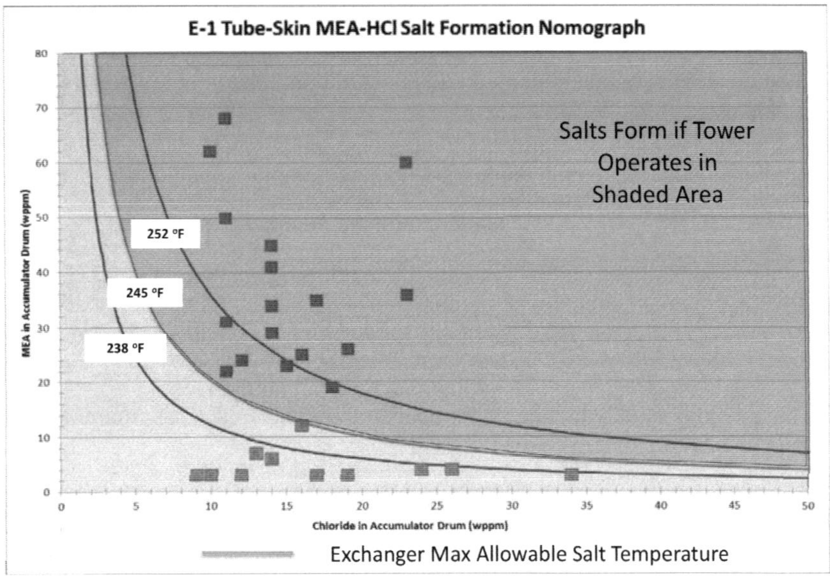

14 AIChE
Spring Meeting, New Orleans, LA

BAKER HUGHES

Shale Oil – Refinery Process Concerns and Monitoring

BAKER HUGHES

CHARACTERISTIC OF SHALE OIL	CONCERN IN REFINERY	PROCESS DATA	SAMPLES AND ANALYSES Samples	 Analysis	OUTPUT DATA	FREQUENCY	BHI PROGRAMS
Crystalline Wax	Tank bottom sludge	• Tank Sludge Depth	• Tank bottom	• Visual sludge • GC analysis	• Tank Sludge Depth • Crystal Wax Presence	• Monthly	• XERIC™ Pretreatment
	Cold Preheat Fouling	• Desalter Temperature			• Temp vs. Time	• Update 2x/mo.	• LIFESPAN™ Wax Dispersants
	Desalter emulsions	• Emulsion layer growth	• Triline emulsions	• Visual wax • GC analysis	• Crystal Wax Presence	• After Shale Oil begins • When emulsion grows	• XERIC Demulsifiers
Asphaltene Stability in Blends	Desalter emulsions	• Emulsion layer growth	• Triline emulsions	• Precipitated Asphaltenes	• Precipitated Asphaltenes	• After Shale Oil beings • When emulsion grows	• XERIC Asphaltene Stabilizers • Field ASIT services™
	ColdPreheat Fouling	• Desalter temperature	• Crude blend	• ASIT	• Temp vs. Time • ASIT #	• Pre-Shale Oil • Update 2x/mo.	• LIFESPAN Antifoulants • Field ASIT services
	Hot Preheat Fouling	• Furnace inlet temp • Furnace tube temps • Furnace fuel usage • Preheat DP	• Crude blend	• ASIT	• Temp vs. Time • Fuel use vs. Time • ASIT #	• Pre-Shale Oil • Update 2x/mo.	• LIFESPAN Antifoulants • MILESTONE™ Furnace Fouling Control • Field ASIT services
Paraffinic	Furnace Coking	• Furnace tube temps • Furnace fuel usage	• Crude blend	• HLPS	• Coking tendency	• Pre-Shale Oil • W/Shale Oil	• MILESTONE Furnace Fouling Control
Tramp MEA	Tower/OH corrosion (Early Warning)	• OH Neutralizer rate • EXCAL usage rate	• OH water • Desalter brine	• OH pH • Brine pH	• Early Sign of Amine • Early sign of Amine	• Pre-Shale Oil • W/Shale Oil	• TOPGUARD™ Overhead Corrosion Control • EXCALIBUR™ Contaminant Removal Technology
	Tower/OH corrosion (Quantitative)	• Inputs for CRM	• OH water • Desalter ww • Desalter brine	• IC Amines	• Confirm MEA in Crude • OH MEA ppm • Salt Risk Tower • Salt Risk OH System	• Pre-Shale Oil • W/Shale Oil • Update 2x/mo.	• TOPGUARD Corrosion Control • EXCALIBUR Technology • TOPGUARD Corrosion Risk Monitor (CRM)
Phosphorus	Tower P Fouling	• Tower DP			• Tower DP Increase	• 2x/mo.	• LIFESPAN Antifoulants, Dispersants
			• Crude	• ICP for P	• P in Crude	• 2x/mo.	
			• Kerosene	• Filt Solids • XRF Filter P • ICP for P	• Incr. Kero Solids • P in Solids • P in Kerosene	• 2x/mo.	
			• Kerosene • Diesel	• ASTM D86	• Incr. in Diesel IPB – Kero EP overlap	• 2x/mo.	

EXCALIBUR, FIELD ASIT SERVICES, LIFESPAN, MILESTONE, TOPGUARD and XERIC are trademarks of Baker Hughes Incorporated.

353965 Hot Oil Heating Systems – A Literature Survey of Engineering Aspects and Applications

Abstract:

A brief technical review and literature survey of the design and operation of hot oil heating systems will be presented. Topics covered will include:

General system design

Engineering considerations

Applications of hot oil systems in the chemical industry

The focus will be to provide the audience an overview of process heating with hot oil and resources/references for those considering this technology.

Heat Transfer Fluid Maintenance for Liquid Phase Systems

Chris Armstrong, Sr Technical Services Advisor

Chris is a Sr. Technical Services Advisor - Specialty and Non-Additized Products for Petro-Canada Lubricants Inc, a Suncor Energy business, based out of Toronto, Canada. He has over 19 years of experience with Petro-Canada in Refinery Operations, Research & Development and Sales. He holds a Materials Engineering Technology Diploma from Mohawk College.

Tel: 1-416-347-3739

charsmtrong@suncor.com

Gaston Arseneault, Sr Technical Services Advisor

Located in New Jersey, Gaston has been a Technical Service Advisor at Suncor Energy/Petro-Canada Lubricants for 14 years, supporting the CALFLO heat transfer fluid product line and managing the related heat transfer fluid testing program. Gaston holds a Masters in Analytical Chemistry from the University of Montreal and authored articles on this subject in *Chemical Engineering*, *Chemical Engineering Progress*, *World Pumps* and *Process Heating*. He holds numerous industry certifications such as the Society of Tribologists and Lubrication Engineers' (STLE) Certified Lubrication Specialist (CLS) and Oil Monitoring Analyst (OMA) I & II, as well as credentials with the International Council for Machinery Lubrication (ICML) and National Lubricating Grease Institute (NLGI - CLGS).

Tel : 1-973-673-3164

garseneault@suncor.com

This presentation focuses on liquid phase systems, which represents the vast majority of heat transfer fluid applications.

1.0 Introduction

Section 1 reviews basic concepts and definitions.

In the context of this presentation which focuses on liquid phase systems, a heat transfer fluid can be defined as:

> Heat Transfer Fluid: A liquid used to carry thermal energy away from the source of heat to a different location on the system, while protecting components and conductors it encounters against rust and corrosion.

There are two designs of liquid phase systems, closed and open systems. Understanding each is very important as it impacts the safety of the operation and the selection of the heat transfer fluid.

In the Open heat transfer systems, the heated fluid comes in direct contact with the atmosphere. In this type of unit, the fluid selected must have an Open Cup flash point (ASTM D92) higher than the operating temperature to ensure a safe operation. In fact, we suggest the fluid flash point be at least 30°F higher than the operating temperature. These systems are generally small and need frequent fluid change because of the high rate of oxidation.

A <u>Closed</u> system is where the <u>heated oil does not come in direct contact with air</u>. These systems have an expansion tank to accommodate the significant volume expansion of the fluid. The expansion tank, also called surge tank, may have air in the headspace. The oil temperature in the expansions tank should never be near the system fluid operating temperature. Many people think that if the expansion reservoir is open to the atmosphere then it must be an open system, hence the possible confusion, which re-emphasizes the need for a clear understanding of what each design means in terms of safety and fluid temperature.

Another concept we wish to explain in this introduction is the Skin Film Temperature (T_{film}) and the Bulk Oil Operating Temperature (T_{bulk}).

The Bulk Oil Temperature is the temperature set by the end user and shown during operation as the heater outlet temperature or heater set point. When looking at a cross-section of a pipe with hot oil circulating in it, as seen in Figure 1, the T_{bulk} is the temperature of the oil in the center of the pipe.

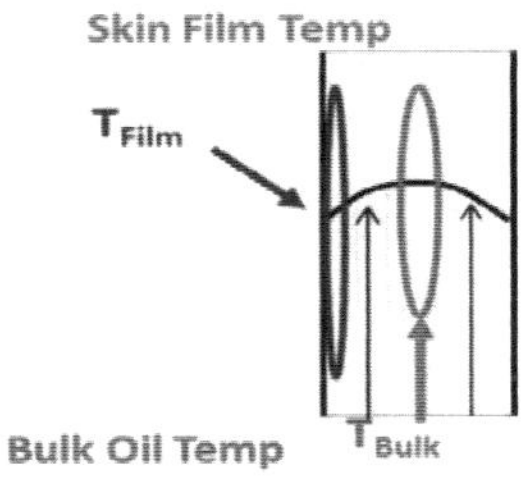

Figure 1: Skin Film and Bulk Oil Temperatures of a heat transfer fluid circulating in a pipe.

Meanwhile, the Skin Film Temperature, also known as the Tubeskin Temperature, reflects the temperature the fluid molecules are exposed to immediately against the pipe wall, the heating element or any surface separating them from the direct heat. In well-designed systems where the flow rate is sufficient, efficient heat transfer occurs (i.e. high Reynolds Number) and the heat flux is reasonable, the T_{film} will be slightly higher than T_{bulk}. However, if a fluid has a high viscosity or the flow rate is reduced due to pump issues or obstruction in the piping, the incremental energy needed to maintain the needed T_{bulk} will cause the T_{film} to be much higher than T_{bulk}. In this situation, a fluid used near its rated maximum bulk oil temperature may suddenly be exposed to significantly higher temperature than it can safely handle and could end up suffering from thermal cracking or operating near its auto-ignition temperature (AIT).

The gap between the T_{film} and T_{bulk} is affected by system parameters (pump sizing, heater conditions, pipe diameter), fluid properties (viscosity, density, thermal conductivity, heat capacity and operating conditions (fluid velocity, raw thermal energy generated by the heat source). In fact, it is preferable to select a fluid based on the application's skin film temperature if it is known, rather than rely on the T_{bulk}. That has proved to be a challenge for systems that have been modified or where original records vanished.

The fluid's ultimate heat transfer performance is characterized by the Heat Transfer Coefficient (*h*), determined by the equation below:

$$\mathbf{h} \sim 0.023\, k^{0.667}\, \mathbf{C_p}^{0.333}\, (V\rho)^{0.8} \,/\, (\,\mu^{0.467}\, D^{0.2})$$

where	k = fluid thermal conductivity	(fluid)	P = fluid density	(fluid)
	$\mathbf{C_p}$ = heat capacity	(fluid)	μ = fluid viscosity	(fluid)
	V = fluid velocity	(operation)	D = pipe diameter	(design)

The fluid ability to transfer heat is affected by some system design parameters as well as its own properties. It is being taught in lubrication classes that lower viscosity oils dissipate heat better than heavier ones, this equation proves why. An increase in viscosity will cause a drop in heat transfer performance. This is applicable when comparing two fluids against one another, but also when dealing with a situation where degradation causes the fluid to increase in viscosity. To put in perspective, an **<u>increase of oil viscosity by 50% will cause a decrease of 17% of the heat transfer coefficient</u>**. This is very significant as it directly affects productivity and profitability.

The last topic in this introduction is concerning the types of heat transfer fluids. Even within a category it is ill advised to generalize but the goal is to raise awareness on the importance of education on the chemistry of the fluid purchased or to be considered. For example, polyglycol synthetics (PAGs) are hygroscopic by nature and absorb water, which can cause pump cavitation issues in some processes. However, we cannot generalize about their Health & Safety because while some carry serious warnings, others are edible and used in products such as toothpaste. Same with mineral based fluids, while some perform poorly and degrade rapidly, others made from severely hydrotreated paraffinic base oils have been classified as non-toxic, inherently biodegradable and superior in thermal stability to their under-performing counterparts. Unfortunately, many individuals do not paint the whole picture about their fluid technology when asked questions by end users, leaving only those with a certain curiosity and extra time to seek the whole truth on their own.

2.0 Degradation of Heat Transfer Fluids

The three (3) primary causes contributing to the degradation of heat transfer fluids are:

1. Thermal degradation (thermal cracking)
2. Oxidative degradation (oxidation)
3. Contamination

2.1 Thermal Degradation (Thermal Cracking)

Thermal degradation of the fluid occurs at the heat source. It happens when the oil molecules receive more thermal energy (heat) then what they can absorb in their short residence time inside the heat source. The excess energy causes thermal cracking (or thermal degradation) of the fluid. It's a function of the oil's ability to absorb the heat but also the heat flux inside the heat source, i.e. the amount of BTUs it receives during its residence time (tied to flow rate).

The result of thermal degradation is seen in Figure 2, showing an example of a typical hydrocarbon molecule found in a mineral based heat transfer fluid undergoing thermal cracking.

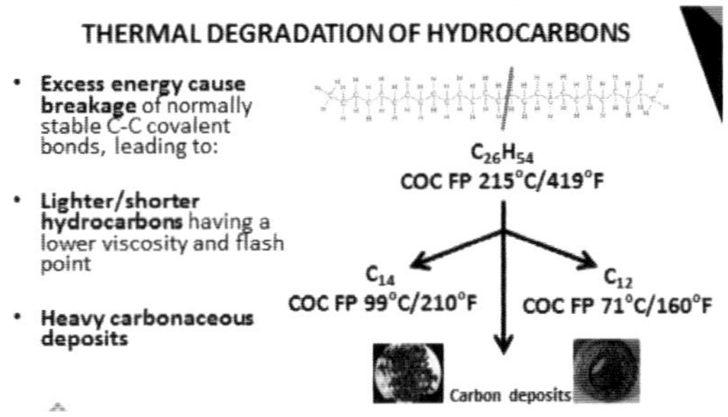

Figure 2: Example of degradation of a mineral based heat transfer fluid

The excess energy breaks the long hydrocarbon molecule (26 carbons long) into 2 shorter molecules of 14 and 12 carbons. These molecules, also called low boilers, have lower flash points. Over time, as their concentration grows, they will contribute to reduce the overall fluid flash and fire points. Secondly, the cleavage reaction forms "high boilers". These are high carbon, low hydrogen molecules that tend to agglomerate together and either bake on the hot surface of the heat source or obstruct lines and elbows downstream on the system. As they collect and bake harder and the carbonaceous residues form a layer that acts as insulator. This creates the vicious circle of thermal degradation (Figure 3).

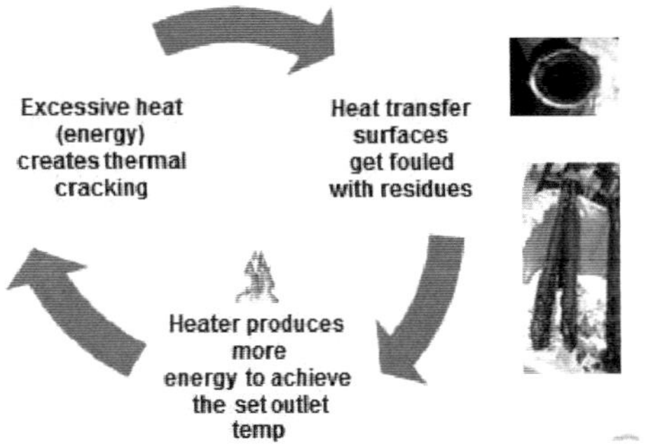

Figure 3: Vicious circle of thermal degradation of heat transfer fluids

In order to maintain the same heater set temperature, the heater now has to produce enough BTUs to not only go through the pipe wall but the additional layer of carbon to heat the fluid. That additional heat raises the T_{film} and the gap between the T_{film} and T_{bulk} widens.

In the field start-up and shut down practices account for many thermal cracking situations. Due to pressure to start operating as fast as possible, system start-up is shortened and the temperature is ramped up too high. Starting a system from ambient room temperature to 200-250°C (400-500°F) should take several hours in order to minimize the heat flux. Shut down can have a destructive effect on the oil if the pump is shut off when the heat is turned off. Furnaces in particular contain refractory material that can hold high heat for several hours, so stagnant oil in the piping will simply undergo thermal cracking. Fluid circulation must be maintained for several hours when the heat is turned off to prevent exposing the fluid to excessive temperatures.

The following tips can help ensure the least amount of energy is used to achieve the desired result:

- Be mindful of the heat flux and control it
- Maintain the pump, heater and fluid in good condition (through oil analysis, PMs, frequent checks)
- Clean the system to restore its efficiency
- Adopt best practices around shut down and start-up

2.2 Fluid Oxidation

Oxidation means reacting with oxygen. Like food or organic matter, heat transfer fluids react with oxygen and degrade. As one can imagine, oxidation is largely related to temperature, the hotter it is the faster oxidation will take place. The rule of thumb is that **oxidation doubles, effectively reducing the life of the oil by 50% with every 10°C/18°F increase.**

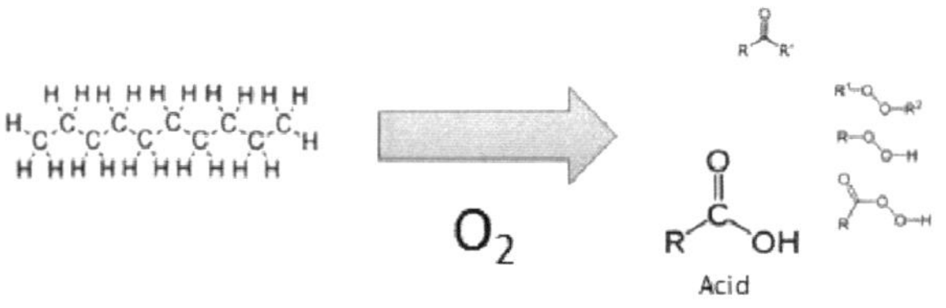

Figure 3: Example of oxidation of a hydrocarbon

Degradation of organic heat transfer fluids, such as mineral based or synthetic chemical aromatics for example, lead to the formation of carboxylic acids, ketones, aldehydes and other substances. Table I below outlines the sequence of events once a fluid starts oxidizing.

Table I: Sequence of events when heat transfer fluids undergo degradation by oxidation

Changes in the fluid	Possible Effect
Gradual discoloration	Minimal
Increase in viscosity	Reduced system performance and efficiency
Increase in acidity	HTF more corrosive towards sensitive metals
Formation of varnish	Valve sticking causing heating issues
Formation of heavy, insoluble compounds	Reduced system efficiency through: • Obstruction of small lines and elbows • Deposits on cool heat exchange surfaces

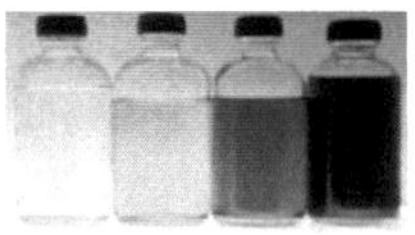

The products of fluid oxidation that settled have fallen out of solution so they are not carried by the oil (solvent) anymore. Therefore, draining the heat transfer fluid will not remove them. At that point, only manual removal and/or utilization of a chemical cleaner to circulate through the system can bring hope to restore the system's initial efficiency. Any residuals of these highly acidic compounds will accelerate corrosion and significantly reduce the life of the next charge of fluid to come in contact with it.

In an open heat transfer system, the hot fluid is constantly in contact with air and there is little that can be done about it. Frequent fluid change-out is required and a system cleaning and flushing is highly recommended every third fluid change.

In a closed system however, the only place where the oil comes in contact with air is in the expansion tank. The fluid temperature in this part of the system should be 60°C/140°F or less, cool enough to leave your hand on it. Oxidation is very slow at this temperature but occurs nonetheless. The best way to eliminate fluid oxidation in a closed system is to install an inert gas blanket in the headspace of the expansion tank, with nitrogen the most popular gas used for this purpose.

2.3 Contamination

This section discussed the potential sources for contamination of heat transfer fluids, classified as Internal and External.

Sources of Internal Contamination:

- Newly commissioned systems present a real threat for contamination. Too often new systems are pressure-tested with water do not get drained completely and end users discovers during a difficult start-up that the new fluid is contaminated with hundreds of gallons of water. New systems also contain welding fines, pieces of wood, metalworking fluid traces (containing sulfur and chlorine) from the machine shop or various other sources. Flushing a new system with a virgin mineral oil or the fluid to be used is always advisable.
- Process leaks are an ongoing threat, as the material being heated can leak into the heat transfer fluid stream through pinholes or cracks. Asphalt terminals know that asphalt leaking in a mineral based heat transfer fluid blends in easily can go unnoticed for months if oil analysis is not performed. The search for the leak point can be expensive and time consuming, let alone fixing it. Chemical plants dealing with strong chemicals know a process leak means immediate shut down, system cleaning and fluid replacement.
- Previous Products. Whatever was ever used in this system can become a contaminant, as it oxidizes and transforms. For Example, previous oils or a cleaner used at one time that was not properly removed

External Contamination refers to the condition of the fluid added and how the oil was stored. For example: Was water sitting on the top of the drum from improper outdoor storage? Is the equipment and containers used to carry the fresh oil to the system dedicated to this fluid only to prevent cross-contamination? These may seem as very obvious observations but we see it all too often even at sites with the best of intentions and safety record.

3.0 Selection Factors (Overall Fluid Ownership Cost)

This section covers the aspects important to consider when selecting a heat transfer fluid.

3.1 Purchase Price

The unit price in dollars per volume or mass is a start but not the end. Small specialty suppliers with no distribution network often ship their products using Less-Than-Truckload (LTL) carriers or quote FOB their warehouse. This adds hundreds of dollars to a purchase. Maintenance people do not see this extra charge because it does not appear on the fluid purchase invoice and the bill goes to the Account Payable department.

3.2 Used Fluid Disposal

Some fluid chemistry's are classified as biodegradable and non-toxic, like hydrotreated petroleum products, which makes disposal easier and less expensive. They are often used a fuel in furnaces if local regulation permits. However, other fluid types can be DOT regulated and even the fresh fluid may contain what OSHA considers Hazardous Material. This will greatly impact the disposal costs and can become a big part of the overall fluid ownership cost.

3.3 Safety/Handling

This is extremely important and often underestimated. Some fluids come with objectionable odors and serious warnings on breathing or potential damage to the reproductive systems if inhaled. Plant personnel in production, laboratories, safety and maintenance need to be fully aware of the safety aspect of handling certain heat transfer fluids. Besides the immediate safety risks you also need to consider the additional cost of compliance to the Material Safety Data Sheet by investing in extra breathing equipment or Personal Protective Equipment (PPE). Training the aforementioned individuals on using and maintaining such equipment adds to the total cost of utilizing a certain fluid in the plant.

3.4 Fluid Change-Out Cycles

If a vendor helps you double the life cycle of the fluid in your application, their fluid price suddenly dropped by 50%. Some vendors promote value while others promote a price. Hence the importance in discussing these talking points. Reducing fluid change-outs also reduces labor, parts and used oil disposal.

3.5 Make-Up Fluid Purchases

Some fluid chemistry's have high vapor pressure, which may cause increased leakage and faster evaporation, requiring more fluid addition over time. It is not uncommon for some chemistry's to require up to 25% of the total system volume in top-up fluid, annually. While this is negligible in a 10 gallons system, it is a large expense in a 10,000 gallons system.

3.6 Fluid Testing

Most fluid vendors are able to offer some form of analysis service. However, only the vendors focused on this heat transfer fluid market offers a true oil testing program using the most relevant test to monitor the health of heat transfer fluids. Some charge hundreds of dollars per sample while others provide it for free.

3.7 Seals Maintenance/Replacement

Some fluid types are known to be inert towards seals, such as mineral oil, while others are more aggressive, requiring more frequent seals replacement. This expense is reflected not only in the seals replacement cost but the weeks and months of fluid leaking prior to the seal being replaced.

4.0 Summary

In conclusion, we hope this paper helps users understand the basic concepts of heat transfer fluids and raise their awareness on the importance of education as the only way to ensure a safe and productive operation, where the system and the fluid's health are monitored closely to provide maximum value.

We discussed in detail how heat transfer fluids can degrade in service, how it's manifested and some ways to reduce or eliminate the threat. Fluid testing is a key element.

Lastly, the case was made to explore all aspects of dealing with heat transfer fluids besides the purchase price. One must consider extra charges for services, oil, parts and labor costs, and most importantly, safety considerations.

This is the blueprint on how to take control of your heat transfer systems and solidify your grip on their operation.

353963 The Multi-Purposes of Heat Transfer Fluid System Expansion Tanks

Abstract:

The use of a heat transfer fluid (hot oil) for process heating will require an adequately sized expansion tank to accommodate the expansion of the heat transfer fluid volume when it is heated to operating temperatures. However, expansion tanks serve additional purposes which can greatly enhance operational safety, equipment life, and the maintainability and life of the heat transfer fluid itself. This presentation will outline the important functions of the expansion tank and how it plays a key role in supporting the effective start-up and operation of high temperature heating systems, so that these key features may be a part of your operating system.

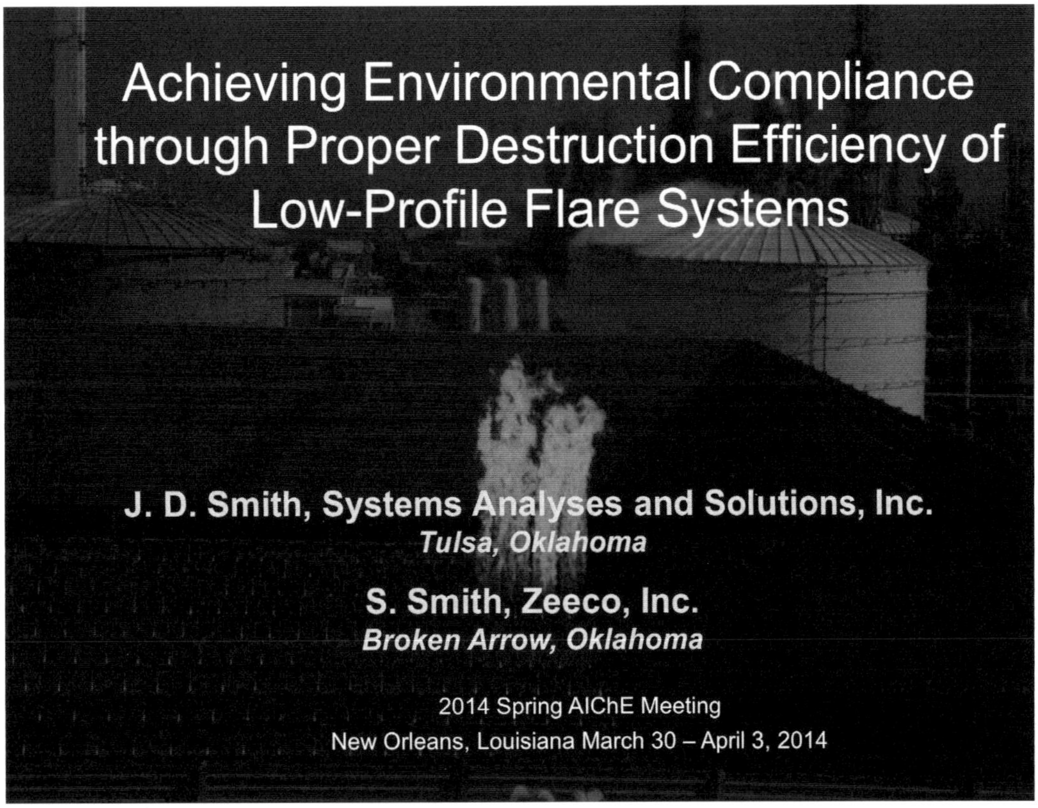
Achieving Environmental Compliance through Proper Destruction Efficiency of Low-Profile Flare Systems
J. D. Smith, Systems Analyses and Solutions, Inc.
Tulsa, Oklahoma
S. Smith, Zeeco, Inc.
Broken Arrow, Oklahoma
2014 Spring AIChE Meeting
New Orleans, Louisiana March 30 – April 3, 2014

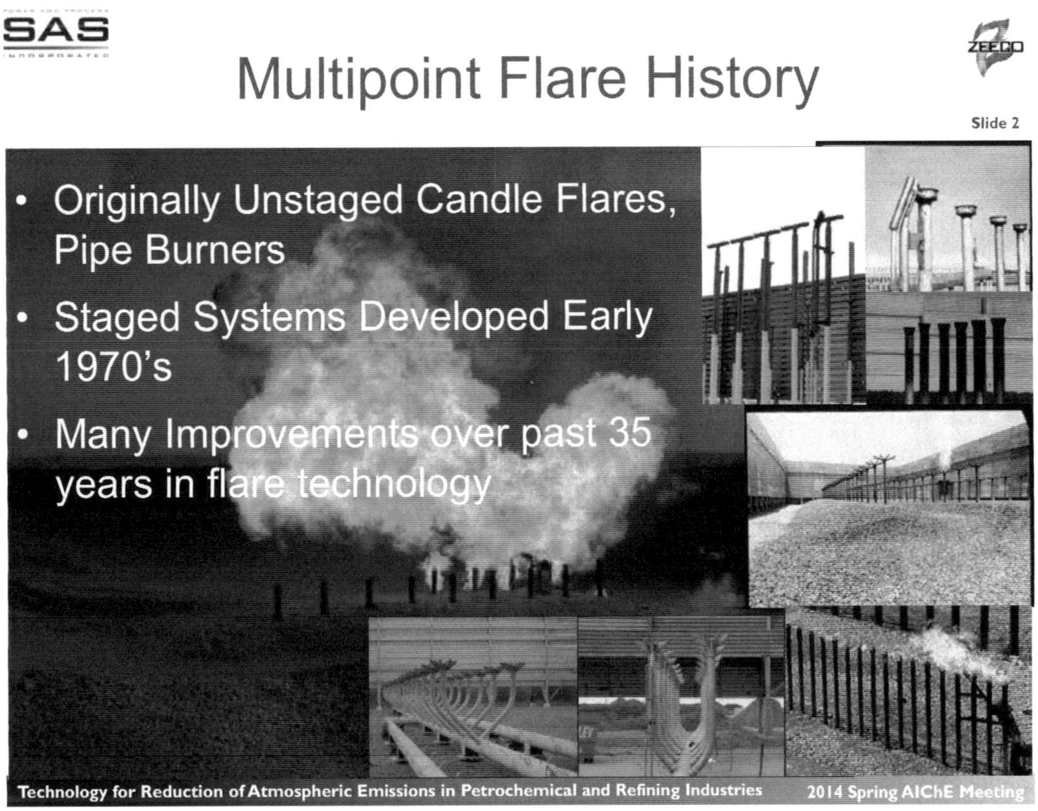
SAS
ZEECO
Multipoint Flare History
Slide 2
• Originally Unstaged Candle Flares, Pipe Burners
• Staged Systems Developed Early 1970's
• Many Improvements over past 35 years in flare technology
Technology for Reduction of Atmospheric Emissions in Petrochemical and Refining Industries
2014 Spring AIChE Meeting

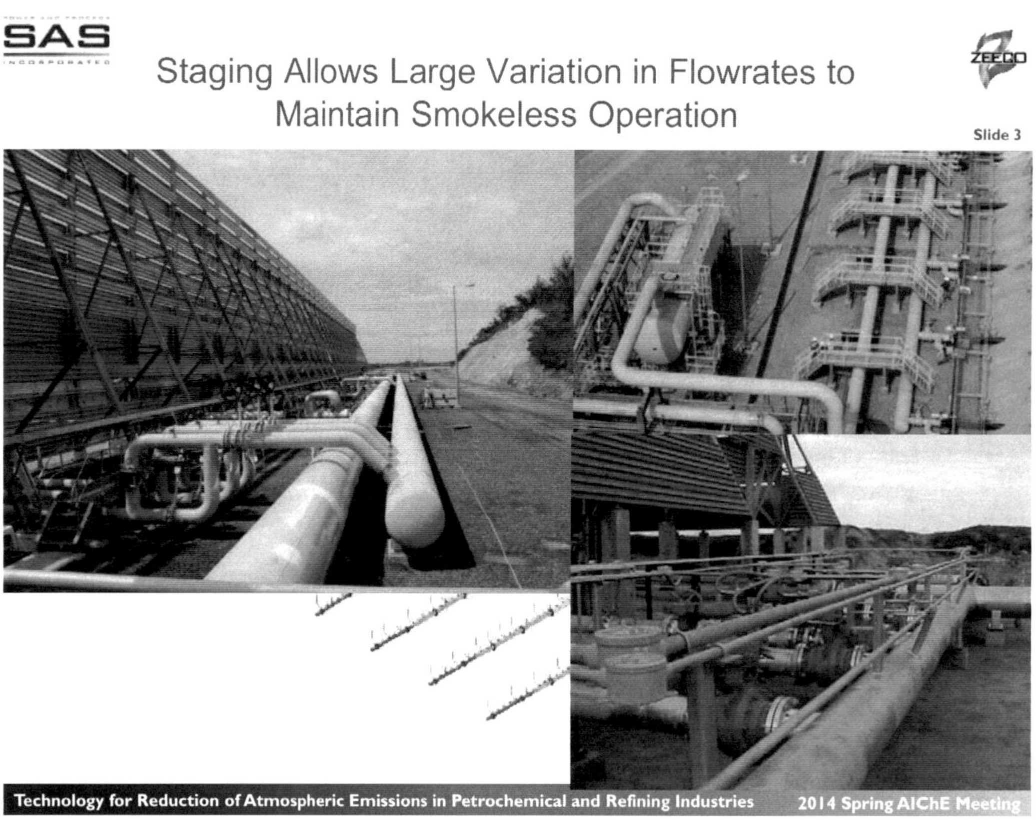
SAS
Staging Allows Large Variation in Flowrates to Maintain Smokeless Operation
ZEECO
Slide 3
Technology for Reduction of Atmospheric Emissions in Petrochemical and Refining Industries
2014 Spring AIChE Meeting

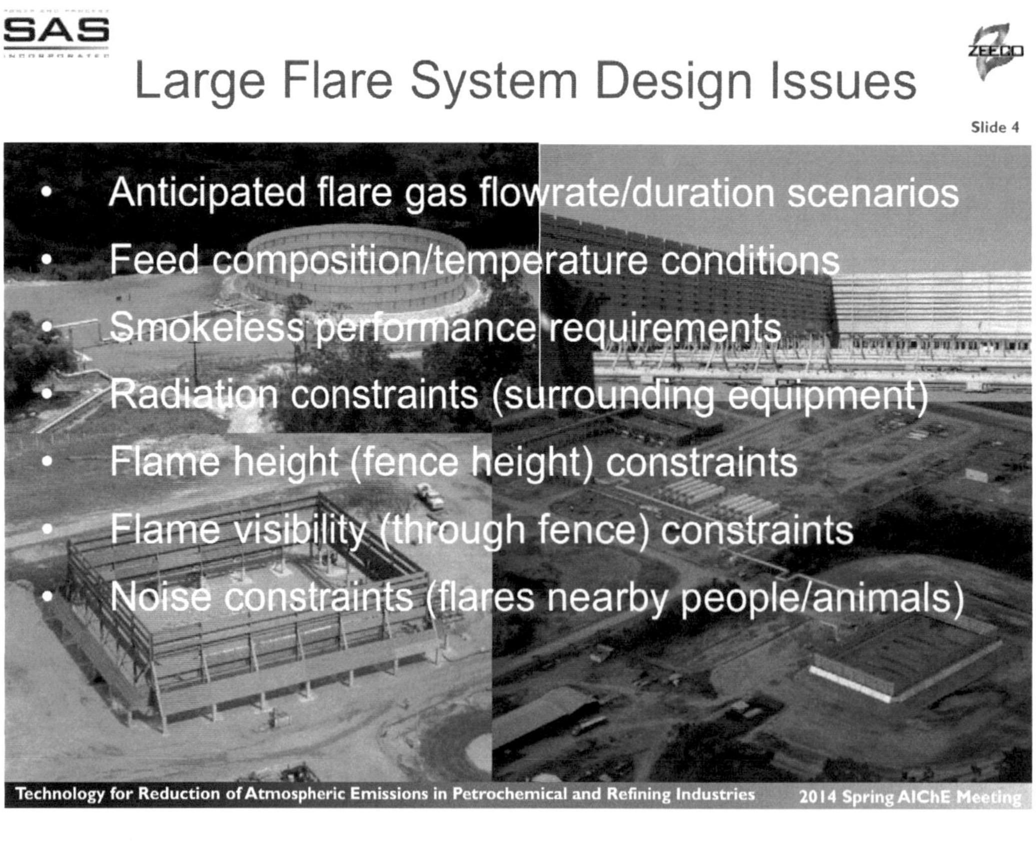
SAS
Large Flare System Design Issues
ZEECO
Slide 4
• Anticipated flare gas flowrate/duration scenarios
• Feed composition/temperature conditions
• Smokeless performance requirements
• Radiation constraints (surrounding equipment)
• Flame height (fence height) constraints
• Flame visibility (through fence) constraints
• Noise constraints (flares nearby people/animals)
Technology for Reduction of Atmospheric Emissions in Petrochemical and Refining Industries
2014 Spring AIChE Meeting

Burner Tip Design based on Multi-arm/ multi-jet configuration

Slide 5

- Variable Arm Area
- Investment Cast
- Pressure Tested at Factory
- 310 SS Cast Material

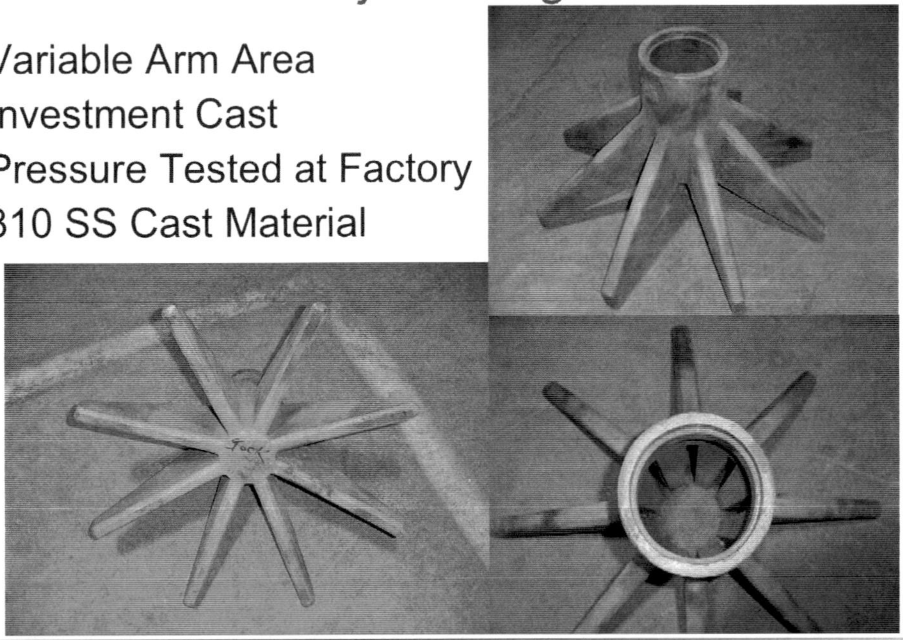

Flare Regulations and Performance Guidelines

Slide 6

- Flaring regulation (API guidelines) focus on utility flares and assisted flares (little attention to multi-tip ground flares)
- Flare Design/Performance guidelines established for general flares – ground flares have additional performance metrics to consider
- Flare performance difficult to quantify (large flame/high radiation flux, elevated operation)
- Multi-Tip ground flares have higher radiation inside large fenced region (larger plumes/difficult to measure emissions!)

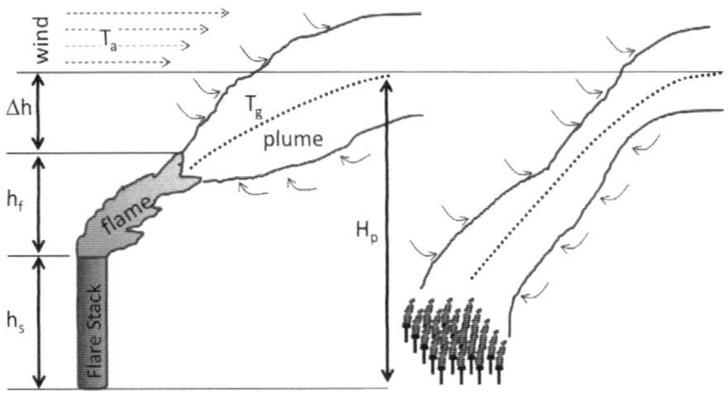

Existing Flare Regulations and Guidelines

Slide 7

- 40 CFR 60.18(b) and 63.11(b)
  - Pilot flame required
  - Sufficiently low exit velocity for stable combustion
  - Heating value high enough for flammable operation
  - Smokeless operation requires adequate mixing energy (air)
- 40 CFR 60.11(d), 60.18(d), 60.482-10(e), 63.6(e), 63.11(b)(1), 63.172(e)
  - Adhere to flare's design documents
  - Use Good Air Pollution Control Practice
- New Source Review

Federal requirements for flares are found in the New Source Performance Standards (NSPS) in § 60.18 and National Emission Standards for Hazardous Air Pollutants (NESHAP) in § 63.11. At a minimum, these rules require flares to be:

- Designed and operated with no visible emissions using EPA Method 22 (except for periods not to exceed 5 minutes in 2 hours);
- Operated with a flame present at all times, confirmed by the use of a thermocouple or equivalent device;
- Used only when the net heating value of the gas to be combusted is 300 BTU per standard cubic foot (BTU/scf) or greater (if the flare is steam- or air-assisted), or 200 BTU/scf or greater (if the flare is nonassisted); and
- Designed for and operated with an exit velocity less than 60 feet per second (ft/sec). An exit velocity of greater than 60 ft/sec but less than 400 ft/sec may be used if the net heating value of the gas being combusted is sufficiently high.

Enforcement Alert, "EPA Enforcement Targets Flaring Efficiency Violoations", EPA-325-F-012-002,August (2012).

Understanding Multi-Tip Ground Flares

Slide 8

- Understanding ground flare performance
  - Where does air come from?
  - How does wind effect flame?
  - How does tip geometry (pressure) affect flame?
- Reacting multi-phase flow in wind
  - Transient flame analysis
  - Near tip mixing
  - 100's of tips in large flare fields
- Model Validation Required
  - Calibrate Smoke and Radiation models

Tip Geometry

Technology for Reduction of Atmospheric Emissions in Petrochemical and Refining Industries 2014 Spring AIChE Meeting

SAS

SAS's C3D Flare Model: Beyond Traditional Flare Analysis

Slide 9

- Extends standard "Steady-State" Flare Analysis:
  - Flame structure interaction with combustion chemistry and radiation flux
  - Fluid interactions with surrounding obstacles
  - Fuel specific soot formation and impact on radiation transfer
- Captures flame dynamics: Momentum vs. Buoyancy, Wind Effect on Flame Structure, Chemical Reactions, Radiation to/from flame
- Assess operating scenarios for specified conditions: wind speed, fuel flow/composition/temperature, impact on/by surroundings
- Key Assumptions:
  - Flame emissivity = F(flare gas composition, soot volume fraction, flame size/shape)
  - Optically thick flames (diffuse radiation) with time dependent view factors
  - Surroundings participate in radiation heat transfer (Re-radiation from ground)
- Overnight analysis on "standard" workstation, animations used to investigate transient flame behavior

Technology for Reduction of Atmospheric Emissions in Petrochemical and Refining Industries 2014 Spring AIChE Meeting

SAS

Multi-Tip Ground Flares: Modeling Issues

Slide 10

- High tip velocity increases air entrainment
  - Tip design critical to air entrainment
  - Local high velocity can translate into high sound levels
- Assist media not available to increase combustion air
  - smoke below certain tip pressure (D-stage pressure)
- Tip spacing critical
  - Flares must cross light
  - Possible Flame merge lengthens flames
  - Adjacent rows compete for air (longer flames, poor performance)

Technology for Reduction of Atmospheric Emissions in Petrochemical and Refining Industries 2014 Spring AIChE Meeting

Multipoint Ground Flare Modeling Approach

Slide 11

Single Burner Test/Model
- Detailed mesh
- Radiation/soot tests
- Single flame height/size

Multi-Burner Test/Model
- Radiation/soot tests
- Flame interactions (spacing)
- Multi-Flame shape/size

Model Full Flare Field
- Use Calibrated Models
- Predict Flare Performance (Smoke/Air Demand)
- Predict Radiation Flux

Transient Flare Analysis: Multi-step Reaction Mechanism

Slide 12

Incomplete Combustion of Flare Gas:

$$1\text{kg } \mathbf{F} + (2.87\text{-}2.6S_1)\text{ kg } \mathbf{O2} \rightarrow S_1\text{ kg } \mathbf{C} + (3.87\text{-}3.6S_1)\text{ kg } \mathbf{PC} + (50\text{-}32S_1)\text{ MJ}$$

Where: (F); Oxygen (O2); Soot (C); Products of Combustion (PC), S_1 soot parameter based on fuel type

Endothermic fuel pyrolysis (cracking reaction – soot producing):

$$1\text{kg } \mathbf{F} + 0.3\text{ MJ} \rightarrow S_2\text{ kg } \mathbf{C} + (1\text{-}S_2)\text{ kg } \mathbf{IS}$$

Where: Intermediate Species (IS), S_2 is cracking parameter based on fuel type

Soot combustion:

$$1\text{kg } \mathbf{C} + 2.6\text{ kg O2} \rightarrow 3.6\text{ kg } \mathbf{CO2} + 32\text{ MJ}$$

Where: Carbon Dioxide (CO2) is combustion product

Intermediate Species combustion:

$$1\text{ kg } IS + \frac{2.87-2.6S_2}{1-S_2}\text{kg}\,O_2 \rightarrow \frac{3.87-3.6\,S_2}{1\text{-}S_2}\text{kg } PC + \frac{50-32S_2}{1\text{-}S_2}MJ$$

Coefficients selected so complete combustion of carbon soot (C) and intermediate species (IS) produce same species and thermal energy as direct combustion of fuel; Coefficients in terms of mass weight (not moles)

SAS INCORPORATED

Flare Burner Tests

Slide 13

Technology for Reduction of Atmospheric Emissions in Petrochemical and Refining Industries 2014 Spring AIChE Meeting

SAS INCORPORATED

Single Tip Burning Propane: wind vs. no-wind

Slide 14

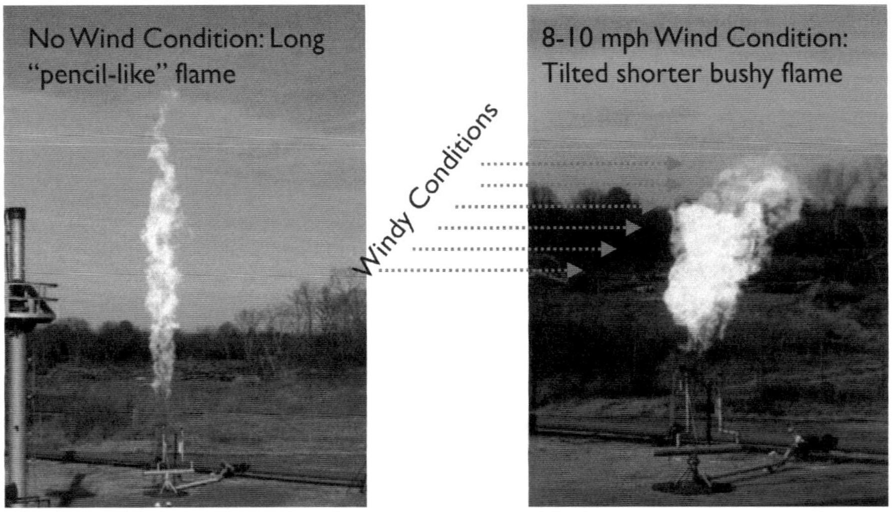

Technology for Reduction of Atmospheric Emissions in Petrochemical and Refining Industries 2014 Spring AIChE Meeting

SAS

Single Burner Flare Model

Slide 15

- 6 X 6 X 26 m physical domain
- Flare Tip located 2 m above ground level
- Turbulence and Arrhenius kinetics Included for fuel gas
  - Reaction Parameters adjusted to match observed flame characteristics
  - Soot Yield matched flame height (i.e., soot burnout)
- Flare Movies for no wind, 3m/s (7mph) wind conditions
- Predicted results for Air demand and Radiation loss from flame determined

Technology for Reduction of Atmospheric Emissions in Petrochemical and Refining Industries 2014 Spring AIChE Meeting

SAS

Combustion Models

Slide 16

Propane:

C_3H_8 + 3.6 O_2 → 3 CO_2 + 1.6 H_2O + 0.024 Soot + 46 MJ/kg propane

Soot + 2.66 O_2 → 3.66 CO_2 + 32 MJ/kg Soot

Ethylene:

C_2H_4 + 0.57 O_2 → 0.93 C_2H_2 + 0.64 H_2O + 9.4 MJ/kg ethylene

C_2H_2 + 2.58 O_2 → 2.7 CO_2 + 0.7 H_2O + 0.2 Soot + 34.1 MJ/kg intermediate

Soot + 2.66 O_2 → 3.66 CO_2 + 32MJ/kg Soot

Mixed Gases:

0.572 C_2H_4 + 0.383 C_2H_6 + 0.043 H_2 +0.982 O_2 →
0.53 C_2H_2 + 0.34 C_2H_3 + 1.1 H_2O + 14.2 MJ/kg

0.61 C_2H_2 + 0.39 C_2H_3 + 2.66 O_2 →
2.66 CO_2 + 0.813 H_2O + 0.181 Soot + 34.4 MJ/kg

Soot + 2.66 O_2 → 3.66 CO_2 + 32 MJ/kg

Technology for Reduction of Atmospheric Emissions in Petrochemical and Refining Industries 2014 Spring AIChE Meeting

Single-burner Mesh

- Rectangular cells
- Local refinement near burner tip
- 110,000 computational cells

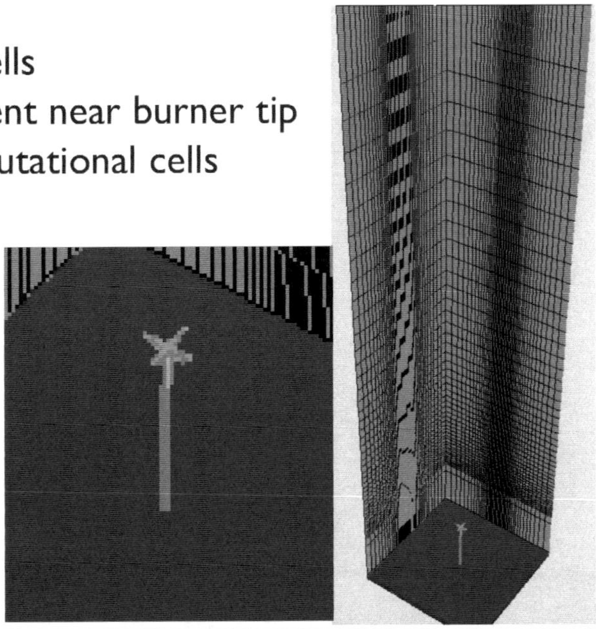

Single Tip Burning Propane: 3.0 m/s Wind

Front View

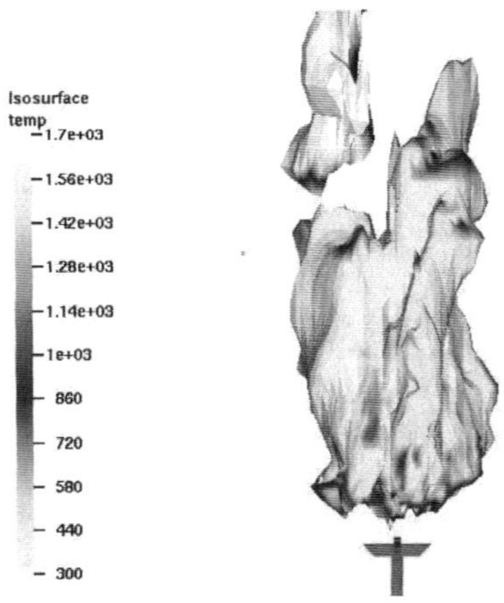

Side View

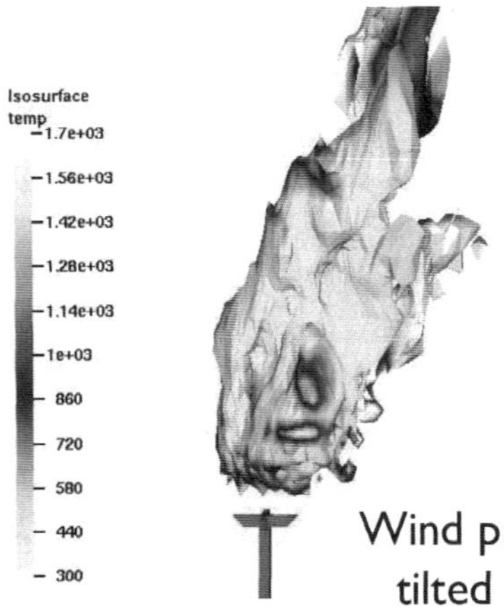

Wind produces tilted bushy, shortened flame

SAS

ZEECO

Single Tip Burning Propane: wind vs. no-wind

Slide 19

Single Burner

No wind produces tight "pencil-like" flames

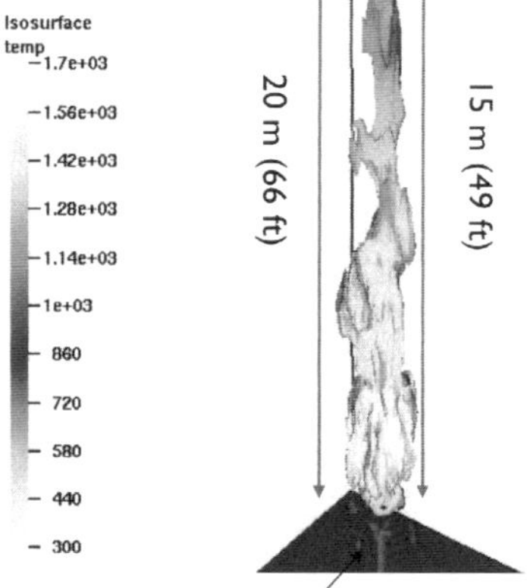

Flare tip located 2 m (~7 ft) above grade

Technology for Reduction of Atmospheric Emissions in Petrochemical and Refining Industries 2014 Spring AIChE Meeting

SAS

ZEECO

Single-Tip Ground Flare Test Results

Slide 20

- 1.4" WC @ 57 °F => 7.3 psig tip pressure
- 12-16 Km/hr (8-10 mph) crosswind
- ~30% flame height reduction
- Minimum flame tilt (~8°)

Technology for Reduction of Atmospheric Emissions in Petrochemical and Refining Industries 2014 Spring AIChE Meeting

SAS

Three-burner Mesh

Slide 21

- Rectangular cells
- Domain size is 30 m X 35 m X 25 m
- Local refinement near burner tips and radiation meters
- 188,000 computational cells

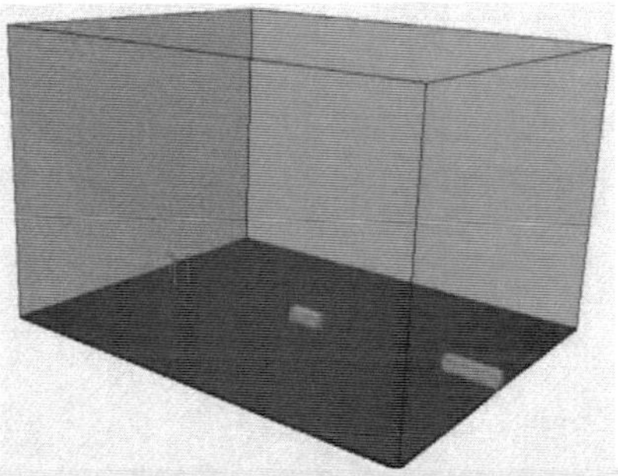

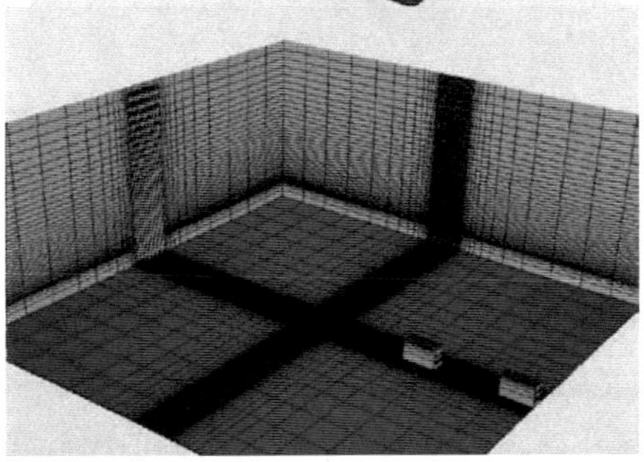

Technology for Reduction of Atmospheric Emissions in Petrochemical and Refining Industries 2014 Spring AIChE Meeting

SAS

Predicted/Measured Flame for 3-burner Ethylene test

Slide 22

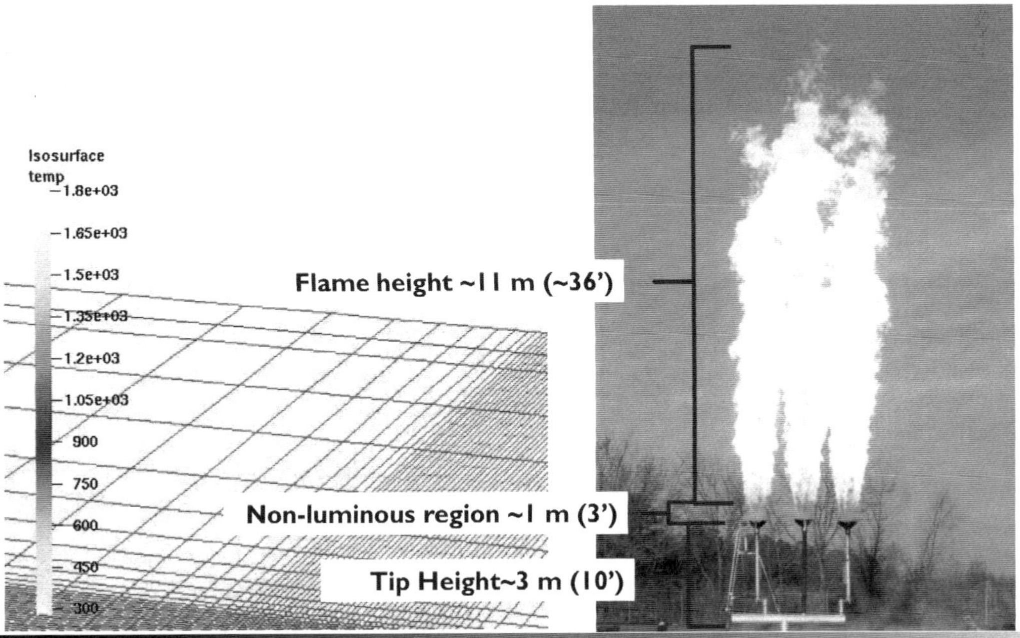

Technology for Reduction of Atmospheric Emissions in Petrochemical and Refining Industries 2014 Spring AIChE Meeting

SAS

Predicted 3-burner flare with radiation monitors

Slide 23

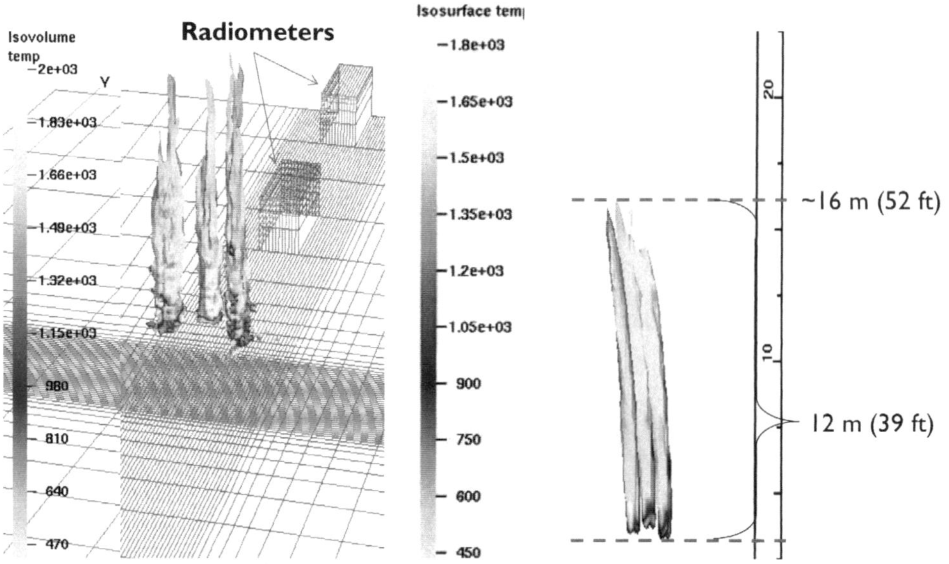

Smith, J.D., Suo-Ahttila, A., Smith, S., and Modi, J., "Estimation of the Air-Demand, Flame Height, and Radiation Load from Low-Profile Flare using ISIS-3D," *American – Japanese Flame Research Committees International Symposium, Advances in Combustion Technology: Improving the Environment and Energy Efficiency*, Marriott Waikoloa, Hawaii - Oct. 22 –24 (2007)

SAS

Predicted Radiation from 3-Burner Flare with Modifications for Ground and Atmospheric Attenuation

Slide 24

Radiation Included in Prediction:

1. Ground re-radiating and reflecting incident radiation from flame
   - Assumed ground $\varepsilon = \alpha = 1$; allow ground to heat to steady state temperature
2. Atmospheric attenuation of radiation from flame
   - Model uses ambient/source temperatures with H_2O/ CO_2 absorption

WALL (plan view perspective)	Left Wall (W/m^2)	Right Wall (W/m^2)	Bottom Wall (W/m^2)	Flame Optical Thickness
PEAK FLOW				
Initial Radiation	78,000	63,000	108,000	0.275
Radiation Modification	61,000	35,000	NA	
SUSTAINED FLOW				
Initial Radiation	15,000	15,000	35,000	0.28
Radiation Modification	6,600	6,600	NA	

SAS

Model Validation: Radiation Predictions Compared to Experimental Data

ZEECO

Slide 25

Tip Size	Position (m)	Burner Pressure (psi)	Total Predicted Radiation (W/m^2)	Measured Radiation (W/m^2)	Difference (%)
3	15	2.8	2700	3344	-20.0 %
3	15	7.3	4750	4803	-1.0 %
3	15	11.4	6150	6192	-0.7 %
3	30	2.8	650	671	-3.0 %
3	30	7.3	1350	1184	+14.0 %
3	30	11.4	1650	1532	+8.0 %
4	15	2.8	4325	6371	-32.0 %
4	15	7.3	8050	8192	-2.0 %
4	15	11.4	10000	9536	+5.0 %
4	30	2.8	1150	1513	-23.0 %
4	30	7.3	2580	2464	+5.0 %
4	30	11.4	3250	2747	+18.0 %

Smith, J.D., Suo-Ahttila, A., Smith, S., and Modi, J., "Estimation of the Air-Demand, Flame Height, and Radiation Load from Low-Profile Flare using ISIS-3D," *American – Japanese Flame Research Committees International Symposium, Advances in Combustion Technology: Improving the Environment and Energy Efficiency*, Marriott Waikoloa, Hawaii - Oct. 22 –24 (2007)

Technology for Reduction of Atmospheric Emissions in Petrochemical and Refining Industries 2014 Spring AIChE Meeting

SAS

Wind Effect on Radiative Flux

ZEECO

Slide 26

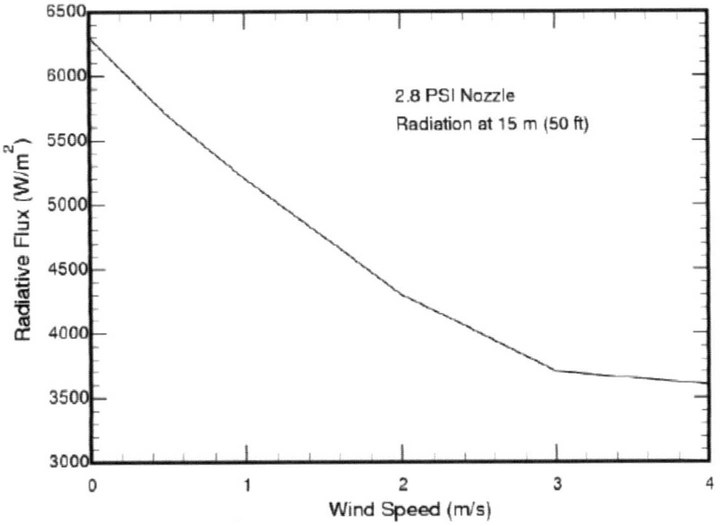

Smith, J.D., Suo-Ahttila, A., Smith, S., and Modi, J., "Estimation of the Air-Demand, Flame Height, and Radiation Load from Low-Profile Flare using ISIS-3D," American – Japanese Flame Research Committees International Symposium, Advances in Combustion Technology: Improving the Environment and Energy Efficiency, Marriott Waikoloa, Hawaii - Oct. 22 –24 (2007)

Technology for Reduction of Atmospheric Emissions in Petrochemical and Refining Industries 2014 Spring AIChE Meeting

SAS

Multipoint Ground Flare

Slide 27

Technology for Reduction of Atmospheric Emissions in Petrochemical and Refining Industries 2014 Spring AIChE Meeting

SAS

Full Field Flare Grid

Slide 28

- Domain size 10 m beyond wind fence; 25 m high
- Local refinement near burner rows/tips
- Sustained Flow Rate: 700,000 kg/hr
- Peak Flow Rate: 1,200,000 kg/hr

Technology for Reduction of Atmospheric Emissions in Petrochemical and Refining Industries 2014 Spring AIChE Meeting

SAS

Simulation of Flare Tests

Slide 29

Predicted flame height well below total fence height

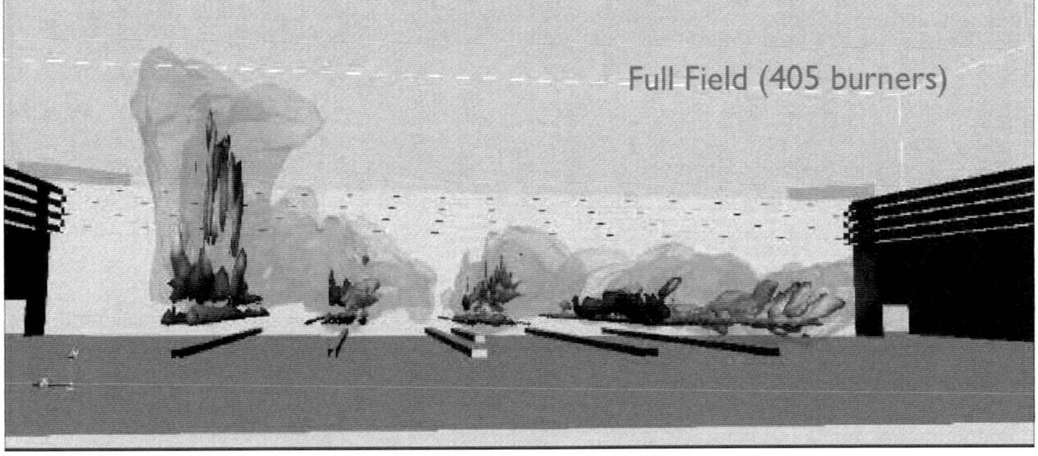

SAS

Max [C_2H_4] along line of sight for peak flow case

Slide 30

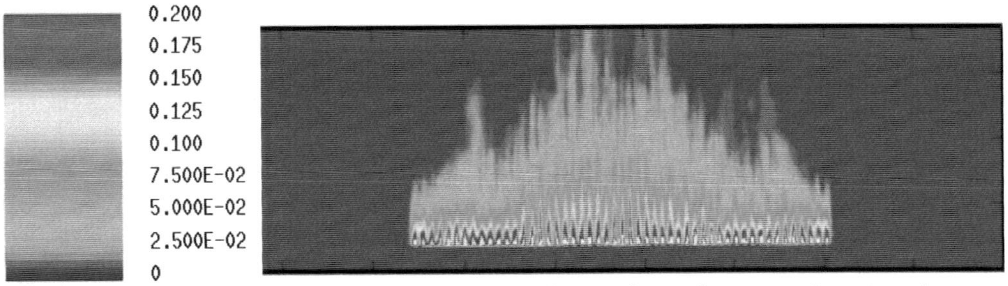

Side view along row showing flame elongation toward center of row

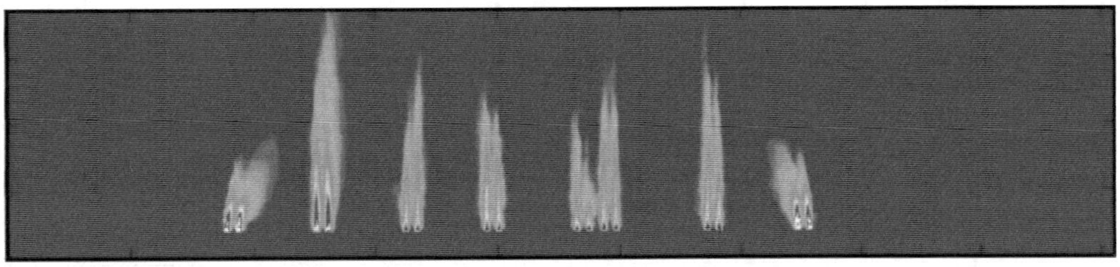

End view of flame height for each row and impact of inflow on outer rows

SAS

Predicted Air/Fuel Requirements for "no-wind" cases

Slide 31

Case Description	Evaluation Plane Area (m^2)	Total Air Flow (kg/s)	Fuel Flow (kg/s)	Air/Fuel Mass Ratio
1 Burner Propane	14.63	60	0.46	130
1 Burner (Tip 3) Ethylene	13.26	52	0.94	55
3 Burner (Tip 3 - 7.3PSI) Ethylene	36	134	2.88	47
3 Burner (Tip 3 - 11.4PSI) Ethylene	36	144	3.84	38
3 Burner (Tip 4 - 7.3PSI) Ethylene	36	150	4.26	35
3 Burner (Tip 4 - 11.4PSI) Ethylene	36	160	5.79	28
Full Field Peak Flow Ethylene	3226	9700	262.3	37
Full Field Sustained Flow Mixed Gas	1843	4800	93.6	51.3

Technology for Reduction of Atmospheric Emissions in Petrochemical and Refining Industries 2014 Spring AIChE Meeting

SAS

Conclusions

Slide 32

- CFD Flare Models:
  - Single-burner model: 110,000 cells
  - Three-burner model: 188,000 cells
  - Full-field model: 700,000 cells (Sustained Flow)/1,200,000 cell (Peak Flow) (Combustion chemistry for propane, ethylene, mixed gas considered)
- Analyzed flame shape/size for 3 fuels for single and three burner tests
- Compared predictions to measurements for 12 tests (2 tip sizes, 3 operating pressures, 2 radiation sample locations)
- "*Reasonable*" estimates of radiation heat transfer and air demand for low profile flare
  - Air/fuel ratios range from 28 to 47 for 3-burner test and from 37 (Peak Flow Case) to 51 (Sustained Flow Case)
- Calibrated flare model applied to full-flare field to estimate:
  - Air demand for specified tip/row spacing
  - Radiation load on wind fence for nominal and peak flow cases
  - Expected flame height and smoke production for nominal and peak flow cases

Technology for Reduction of Atmospheric Emissions in Petrochemical and Refining Industries 2014 Spring AIChE Meeting

SAS

Questions?

Slide 33

353103 A Multi-Period Process Integration to the Management of Process Flares with Consideration of Fuel Substitution and Cogeneration Systems

Abstract:

A wide variety of process upsets can lead to flaring. Such flaring activities cause environmental emissions and result in economic losses. A possible approach to mitigating these emissions and losses is to utilize the flared gases in combustion systems (boilers and industrial furnaces) that can be used to produce utilities such as steam. Additionally, the generated steam can be further used in a cogeneration system (e.g., through steam turbines) to produce power and heating. This paper introduces a process integration approach to the effective utilization of flared gases into combustion and cogeneration systems. Since the flared gases differ in composition and characteristics from the fuel that is typically used in boilers, there is a need to make sure that the fuel quality is in the acceptable range for the boiler. Various constraints may be used to assess usability of flared gases as part of the fuel blend fed to the boiler. For instance, the Wobbe index (WI) may be used to study fuel properties sent to the boilers. A multi-period optimization formulation is developed to determine the optimal usage of flare gases in the boiler and cogeneration systems. An operational period is discretized into multiple periods corresponding to different operating modes of the process (process episodes). Economic objectives are used while including technical and environmental constraints. Additionally, the cost-benefit analysis for adding a cogeneration system with an optimal size is carried out. A case study using an ethylene plant is used to demonstrate the various aspects of the process.

PROOF OF CONCEPT TEST FOR A REAL-TIME FLARE COMBUSTION EFFICIENCY MONITOR

Presented at
2014 AIChE Spring Meeting &
10th Global Congress on Process Safety

By
Yousheng Zeng, PhD, PE and Jon Morris
Providence Photonics, LLC

Mark Dombrowski
Surface Optics Corporation

April 2, 2014

PROVIDENCE

A BRIEF REVIEW OF THE CONCEPT PRESENTED AT 2012 FALL CONFERENCE

Ref.: Zeng and Morris, "A New Method to Measure Flare Combustion Efficiency in Real-Time", presented at AWMA Louisiana Section 2012 Fall Conference, Baton Rouge, Louisiana, October 30-31, 2012

2

PROVIDENCE

Flare Combustion Efficiency (CE):

$$CE(\%) = \frac{[C]_{CO2}}{\sum_i n_i[C]_{HCi}+[C]_{CO2}+[C]_{CO}} \quad \text{Eq. (1)}$$

Flare CE – Very difficult to measure

Source: TCEQ/UT

THE CONCEPT

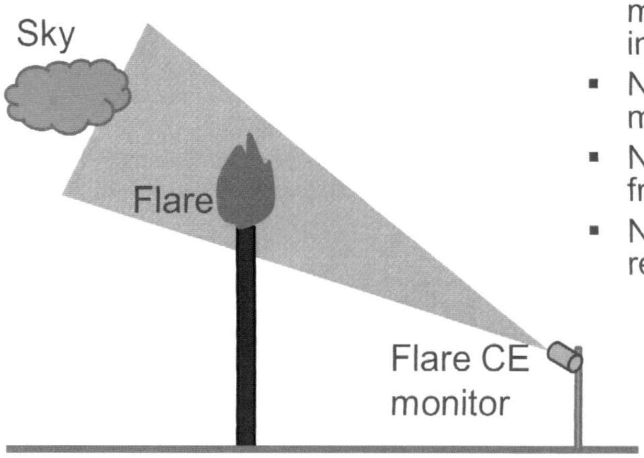

- Monitor and map flare CE in real time through a special multi-spectral IR imaging device
- Not a path measurement
- No scanning; high frame rate
- No operator required

We call it "Flare Efficiency Monitoring System" or "FEMS"

VISION

- Monitor flare CE in real-time
- Image the full flare flame; measure both overall CE and CE at a pixel level (CE mapping)
- Industrial grade device, suitable for integration with PLC or DCS
- One monitor covers multiple flares on site (step and stare)

5

PROVIDENCE

WORKING PRINCIPLE

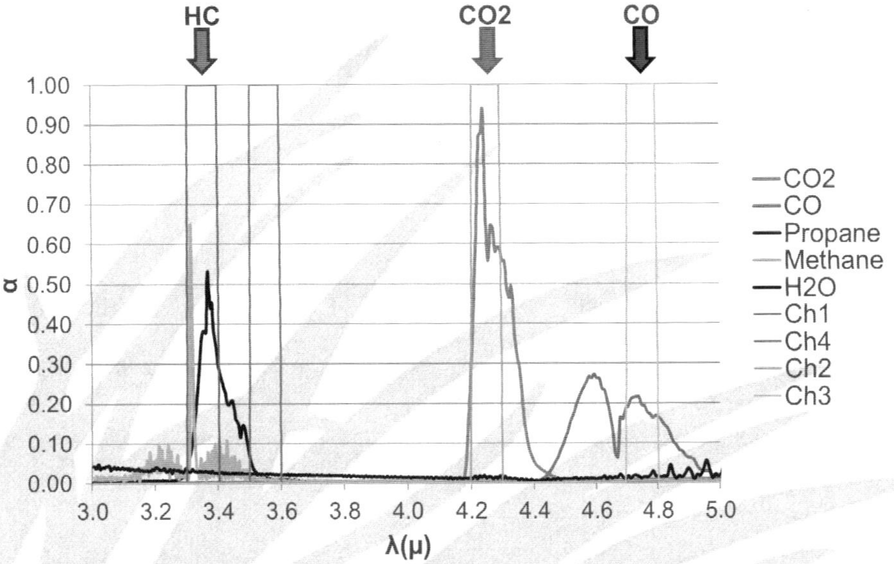

6

PROVIDENCE

PROOF OF CONCEPT TEST

The proof of concept test was partially funded by EPA SBIR Phase I grant.

TEST SETUP

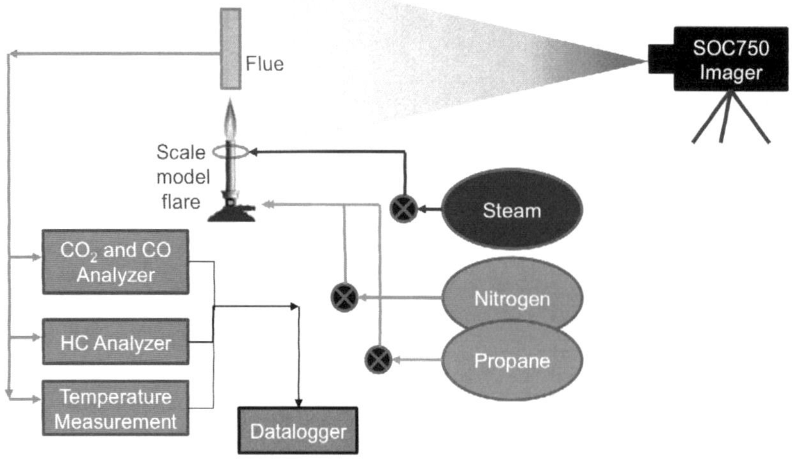

SOC750
Hyper-Spectral
Imager

Scale model flare,
Flue w/ Sampling Probe, and
Analyzers

PROVIDENCE

SCALE MODEL FLARE

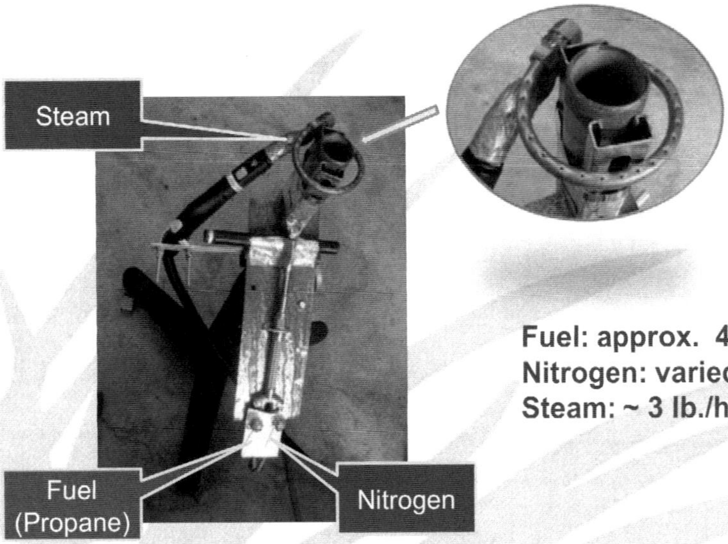

Fuel: approx. 40k Btu/hr.
Nitrogen: varied
Steam: ~ 3 lb./hr.

PROVIDENCE

FLUE AND SAMPLING PROBE

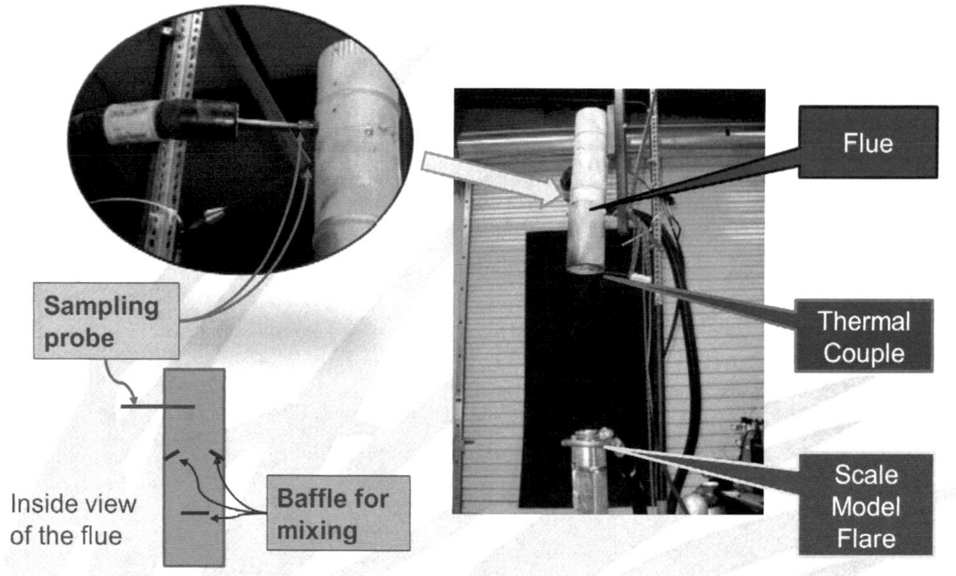

11

PROVIDENCE

ANALYZERS AND IMAGER

Testo 350 XL
(CO2 using NDIR, CO, O2, NOx, H2, Temp, etc.)

3010 Mini FID
Calibrated to Propane

SOC750 Hyper-spectral imager
42 spectral channels, operated at 22 cubes per sec.

12

PROVIDENCE

TEST RUNS

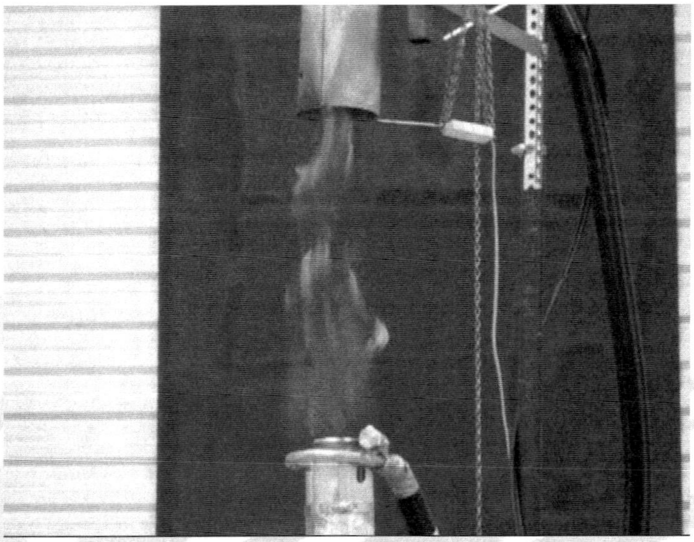

13

PROVIDENCE

PRELIMINARY RESULTS - WITHOUT CALIBRATION

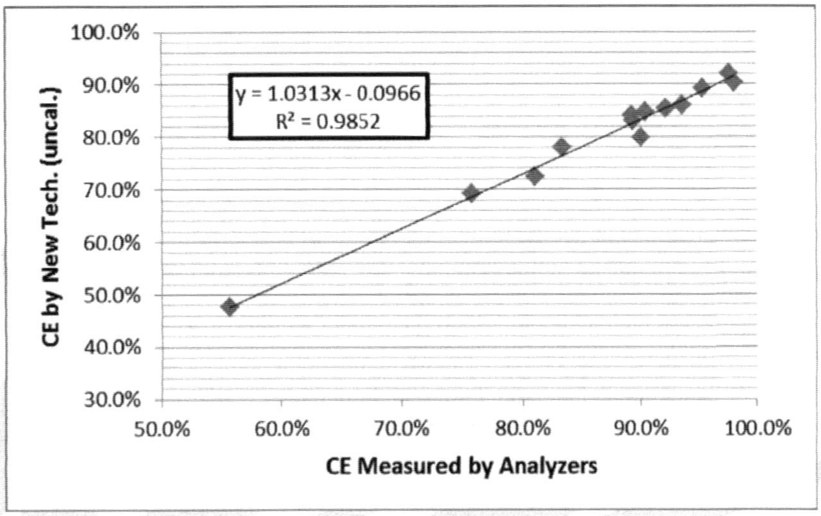

14

PROVIDENCE

Preliminary Results
- with an Initial Calibration

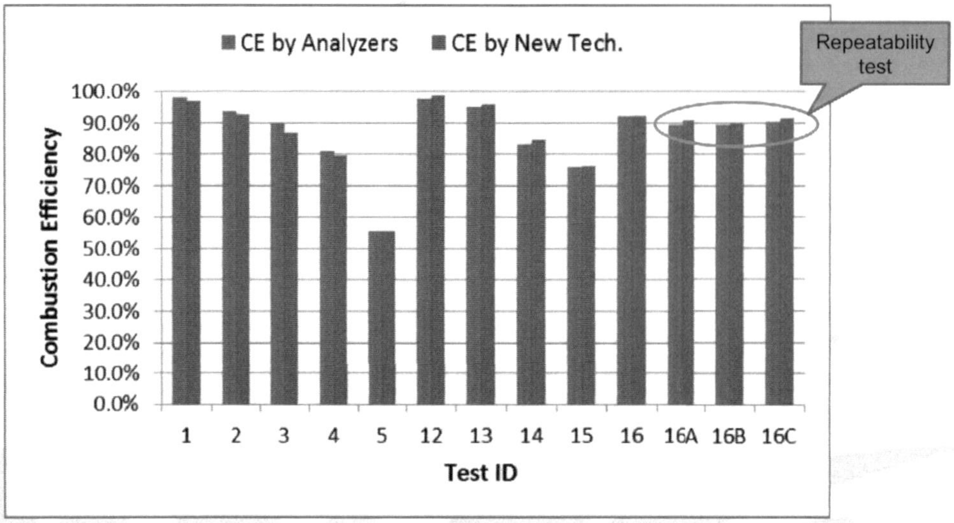

15

PROVIDENCE

Preliminary Results
- Initial Method to Measure Temperature

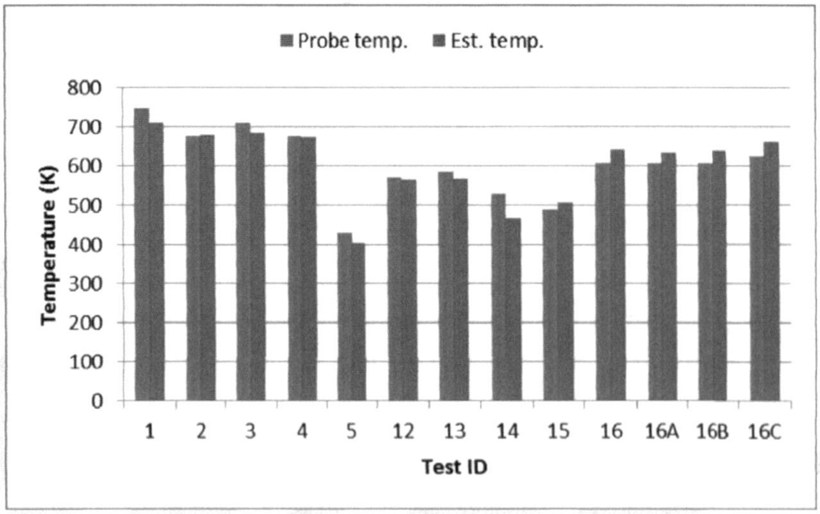

16

PROVIDENCE

PRELIMINARY RESULTS – CE MAP

30 sec. average (670 frames)

1 frame (~0.045 sec.)

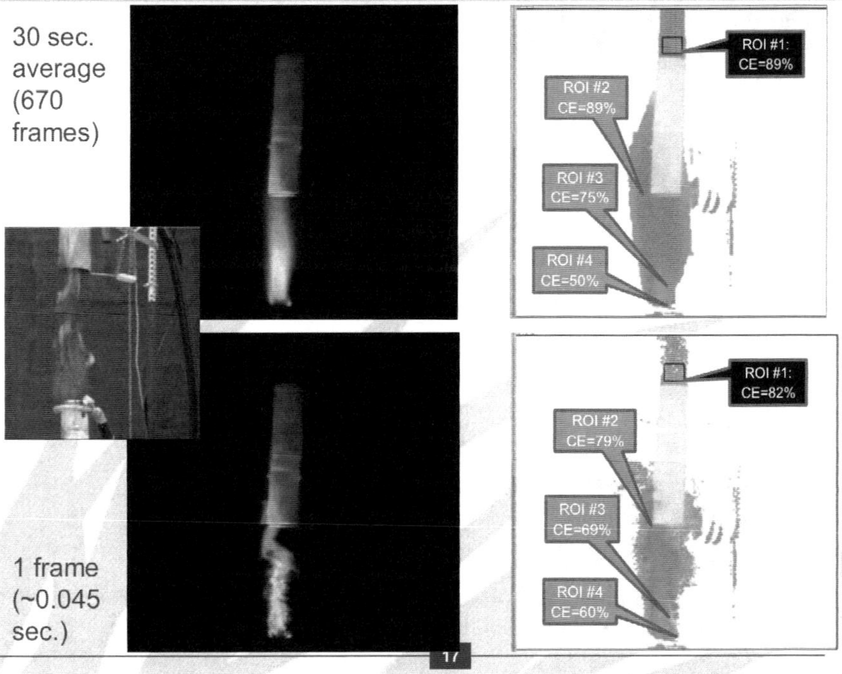

17

Three Types of Measurement

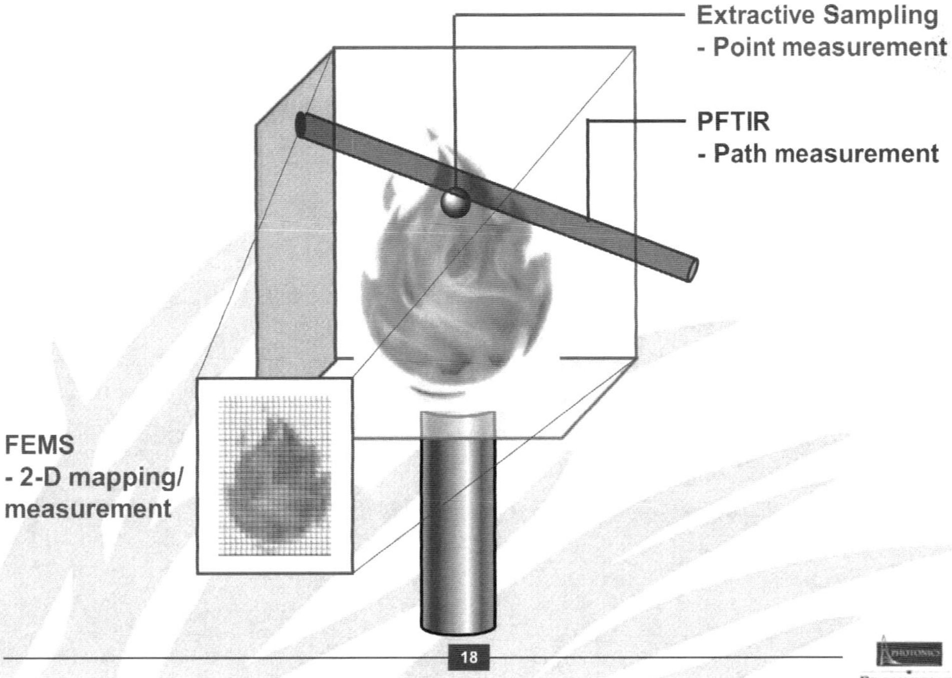

18

Comparison with PFTIR

PFTIR

- "Scanning" - >1 sec/scan – assuming that flare is static during that time
- Path measurement – aiming required
- Human operation

FEMS

- Staring – 20-30 data cubes/sec – match the flare dynamics
- 2-D mapping of CE – no aiming required
- Automatic

19

PHOTONICS PROVIDENCE

Conclusion

- The CE determined by the new technology correlate well with the CE measured by conventional analyzers
- With a further developed calibration method, real-time CE monitoring and feedback for flare optimization is feasible
- The new technology can determine CE at a pixel level, generating a CE map for the entire flare flame. No aiming issue.
- As a side benefit, it can also provide temperature mapping of the flare flame

20

PHOTONICS PROVIDENCE

LOOKING FORWARD…

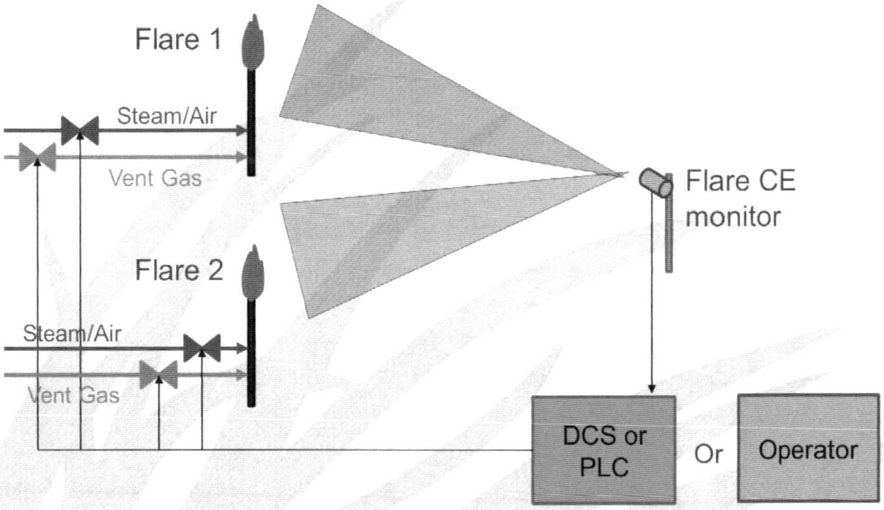

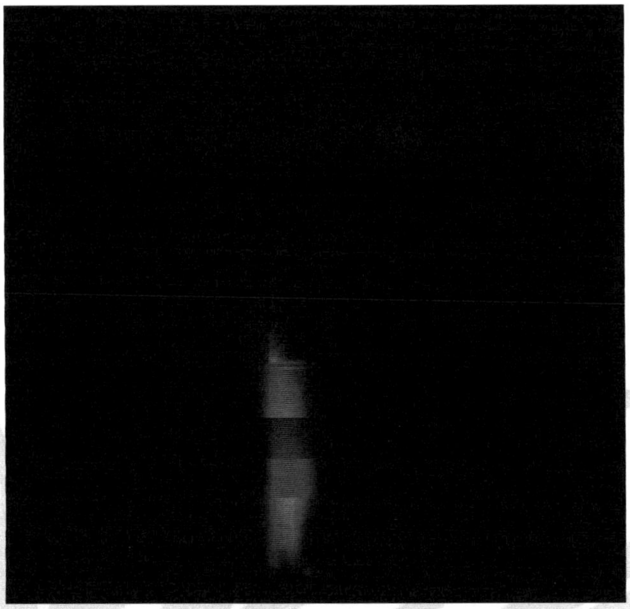

Contact Info. youshengzeng@providenceeng.com

348171 Flameless Thermal Oxidation

Abstract:

Linde's patented Flameless Thermal Oxidizers (FTO) technology oxidizes combustible gases and atomized liquids with air under controlled conditions in a specially designed reactor without a flame. Residence time and reaction temperatures are carefully selected to achieve 99.99 +% destruction efficiency with ultra-low NOx and dioxin emissions at very high thermal energy efficiency. These systems are commercially proven, cost effective and highly effective for the destruction of volatile organic compounds (VOC's) and hazardous air pollutants (HAP's). This paper will present an overview of the FTO technology and market applications with special emphasis on environmental performance and cost considerations.

350916 Modeling of the Dynamic Interaction in an LDPE Closed System Purge Bins-Regenerative Thermal Oxidizer (RTO) for Ethylene Emission Mitigation

Abstract:

Common designs of LDPE purge bins involve venting the purge and blend air to atmosphere containing small traces of ethylene. To mitigate ethylene emission from these bins, a new process design has been proposed to collect vent air-carrying ethylene via a closed ducting system connected to a regenerative thermal oxidizer (RTO) unit. This necessitates a blower at the downstream end to pull the purge air before feeding into the RTO to incinerate the ethylene. Concerns have been raised regarding the impact of such system on the purge bins, which are not designed to withstand any significant amount of vacuum. While conceptually steady operation of such system is feasible, operation of purge bins is never steady particularly when the purge bins are not designed to withstand any significant level of vacuum pressure. With the minimum safe operating pressure at -0.1 psig (-0.68 kPag), and the design RTO inlet pressure being -1"'WC (-0.25 kPag), the bins are forced to operate under slight vacuum conditions with a very narrow allowable margin for error. The operation is inherently dynamic as LDPE pellets are being conveyed into and out of the bins, purged and mixed in cycle by the interaction of multiple pieces of equipment. In the event of any equipment failure along the train, pressure in the bins can drop below their minimum vacuum level and cause damage to the shell. Dynamic interactions between the various components of the system including the blower with its VFD drive, purge and blend air compressors, capacitances and resistive characteristics of the pellets in the bins, as well as the response of the control system all contribute to interesting dynamic interactions that must be attended to. A model capable of simulating various dynamic phenomena has been built using Aspen HYSYS Dynamics TM v7.3 for the entire system, and used to answer what-if operational scenarios. As an example it was shown that the addition of a pull blower with its variable frequency drive (VFD), cannot maintain the bin pressure above the vacuum pressure limit of -0.1 psig even with the installation of individual pressure vacuum relief valve (PVRV) on each bin. Any interruption to air flows such as the shut-down of blend blower(s), stop/start conveying or swinging the purge air blower poses a non-negligible risk. Interaction between the PVRV's and the blower's VFD both simultaneously trying to control the bin pressure could lead to pressure fluctuations that exceed the design vacuum pressure of the bins. This paper will provide an overview of the system, state the design and operational challenges, basis of the dynamic model, and results of several what-if scenarios.

351659 Air Quality Impact of the Startup of a Single Olefin Plant

Abstract:

Ground-level ozone is one of six common pollutants defined by US EPA (Environmental Protection Agency). It can cause health effects, including lung diseases. HRVOC (Highly Reactive Volatile Organic Compounds) emissions from industries are highly responsible for ground-level ozone pollution, especially around industrial district. Flare emissions from MSS (maintenance, startup and shutdown) & events are major sources of HRVOC emissions.

Most of air quality modeling treats industrial flares as constant emissions, which means the emissions keep constant for the whole scenario days, and the total amount of emissions equal to the total emissions reported to environmental authorities. However, real flare emissions from MSS and events are time-various. At certain time, flare emissions could be very huge, and at other time, flare emissions could be very small. During emission peaks, the ozone formation could be much higher than a flat emission curve.

In this paper, time-various flare emissions of startup procedures are considered for air quality modeling. Emission data are obtained from dynamic simulation of a typical ethylene plant. The dynamic simulation results were validated with real plant data and emission data of startup events. The time-various flare emission data were added into Houston-Galveston-Brazoria (HGB) 2006 episode base case modeling as a point emission source. Three startup scenarios were simulated, which include a baseline scenario and two flare reduction scenarios. For each scenario, 120 cases were built to simulate different starting time covering the whole episode days.

Simulation results show that for the baseline scenario, the maximum 8-hour average ozone increment could be 15 ppb at certain starting time. It means starting a single plant could temporary increase the regional ozone concentration by

15 ppb, which is 20% of current ozone standard (75 ppb). For the two flare reduction scenarios, the maximum 8-hour average ozone increment could also reach 10 ppb. Simulation results also show that ozone increment is high when starting time is before 8 am. Ozone increment is low when starting time is after 12 pm. For each scenario, air quality simulation gives least impact period, which means ozone impact could be low when starting time is within that period. For the baseline scenario, the least impact period is only 2 hours. For the two flare reduction scenarios, the least impact periods are 4 hours and 8 hours.

In conclusion, dynamic simulation and air quality simulation show that the flare emissions from MSS and events are highly time-various, and may not be treated as flat emissions. Air quality simulations also give least impact periods of startup time, which could help chemical plants to arrange their startup schedule to reduce ozone formation.

SOLID FUEL TO NATURAL GAS CONVERSIONS FOR EXISTING BOILER APPLICATIONS

William Gurski
Director Power Sales
Zeeco Inc.

John Guarco
Technical Director Boiler Burners
Zeeco Inc.

Nando Nunziante
Manager Power Controls
Zeeco Inc.

Prepared for Presentation at the Year 2014 Spring National Meeting
New Orleans, LA, March 30 – April 3, 2014

Abstract

Recent discoveries of vast natural gas reserves in the United States have resulted in increased production of Natural Gas, resulting in lower prices. Petrochemical facilities are performing solid fuel conversions on their boilers to natural gas as a cost-effective and efficient fuel solution. Natural Gas is not only economically beneficial but also environmentally efficient with cheaper prices and reduced SO_2, NO_x, and CO_2 emissions. As more facilities consider boiler fuel conversions, it is important to understand all facets of the conversion, from the thermal evaluation of the boiler, to the complete design, supply and installation of the new firing system.

We will provide specific details and recommended practices from a recent Circulating Fluidized Bed (CFB) Boiler fuel conversion application designed for 1.4 billion Btu/hr of heat input for the maximum continuous steam rating. We will also review thermal performance results, design and supply of the complete new gas firing system, and installation conversion assistance for the boiler modifications and firing system installation details.

Introduction

Zeeco was contacted by a large industrial plant owner to assist them with the fuel changeover of a utility designed solid fuel CFB. Due to current contract requirements and the competitive nature of the industry, the plant owner will remain confidential.

The primary goals of the project are as follows:

- Full Conversion from Petcoke firing to Natural Gas
- Maintain boiler steam design capacity (~1200 kpph)
- Meet permitted NO_x requirements – 0.07 lbs/MMBtu
    - 25-100% heat input for NO_x emissions
- Minimal or no impact on the boiler waterside design

The owner provided the original boiler design information for Zeeco's review of potential boiler burner retrofitting options. This was the first of two boilers that would require retrofitting Two sister units were built/commissioned on solid fuels approximately three years previously. The units were originally designed with natural gas start-up burners and commissioned for main fuel support via the petcoke. The petcoke was to be fired and combusted via a circulating fluid bed system. Upon initial advisement from the plant owner, they noted that their petcoke fuel contract was no longer cost effective, as this fuel feedstock was significantly higher in price than originally estimated in the plant system's budget. The boilers were only in operation for less than two years when this decision to switch to natural gas was determined.

Zeeco will review the major facets of this boiler conversion—from the initial proposal phase to the commissioning and startup of the boiler and firing system. We will also discuss the combustion parameters of the technology behind the project's success–Zeeco's DT Ultra-Low NO_x Free-Jet Burner technology.

Combustion Parameters

In order to understand why the ZEECO® Free-Jet technology was chosen, formation of thermal NO_x emissions must first be examined. For gaseous fuels with no fuel-bound nitrogen (N_2), thermal NO_x is the primary contributor to overall NO_x production. Thermal NO_x is produced when flame temperatures reach a high enough level to "break" the covalent N_2 bond apart, allowing the "free" nitrogen atoms to bond with oxygen to form NO_x.

Stoichiometric Equation describing typical combustion in a natural gas fired burner:

Methane & Air with Excess Air

$$2CH_4 + 4\ (XA)\ O_2 + 15\ (XR)\ N_2 \text{ ---> } 2CO_2 + 4H_2O + (XA)\ 15N_2 + (XR)\ O_2$$

Natural air is comprised of 21% O_2 and 79% N_2. Combustion occurs when O_2 reacts and combines with fuel (typically hydrocarbon). However, the temperature of combustion is not normally high enough to break all of the N_2 bonds, so a majority of nitrogen in the air stream passes through the combustion process and remains diatomic nitrogen (N_2) in the inert combustion products. Very little N_2 is able to reach high enough temperatures in the high intensity regions of the flame to break apart and form "free" nitrogen. Once the covalent nitrogen bond is broken, the "free" nitrogen is available to bond with other atoms. Basic chemistry dictates that free nitrogen, or nitrogen radicals will react to other atoms or molecules that can accept them to create a more stable atom. Of the possible reactions with the products of combustion, free nitrogen will most likely bond with other free nitrogen to form N_2. However, if a free nitrogen atom is not available, the free nitrogen will react with the oxygen atoms to form thermal NO_x. As the flame temperature increases, the stability of the N_2 covalent bond decreases, allowing the formation of free nitrogen and subsequently increasing thermal NO_x. Burner designers can reduce overall NO_x emissions by decreasing the peak flame temperature, which can reduce the formation of free nitrogen available to form thermal NO_x.

The varied requirements of Power and Industrial plants entail the use of numerous types and configurations of burners. The method utilized to lower NO_x emissions can differ by application. However, thermal NO_x reduction is generally achieved by delaying the rate of combustion. Since the combustion process is a reaction between oxygen and fuel, the objective of delayed combustion is to reduce the rate at which the fuel and oxygen mix together and burn. The faster the oxygen and the fuel gas mix, the faster the rate of combustion and the higher the peak flame temperature.

Figure 1 plots Peak Flame Temperature against Thermal NO_x created. NO_x emissions increase as the adiabatic flame temperature increases. Slowing the combustion reaction reduces the flame temperature, which results in lower thermal NO_x emissions. The challenge in achieving lower thermal NO_x emissions is not the theory; however, it is in retrofitting the latest burner technologies into older existing boilers without adding expensive external components or processes.

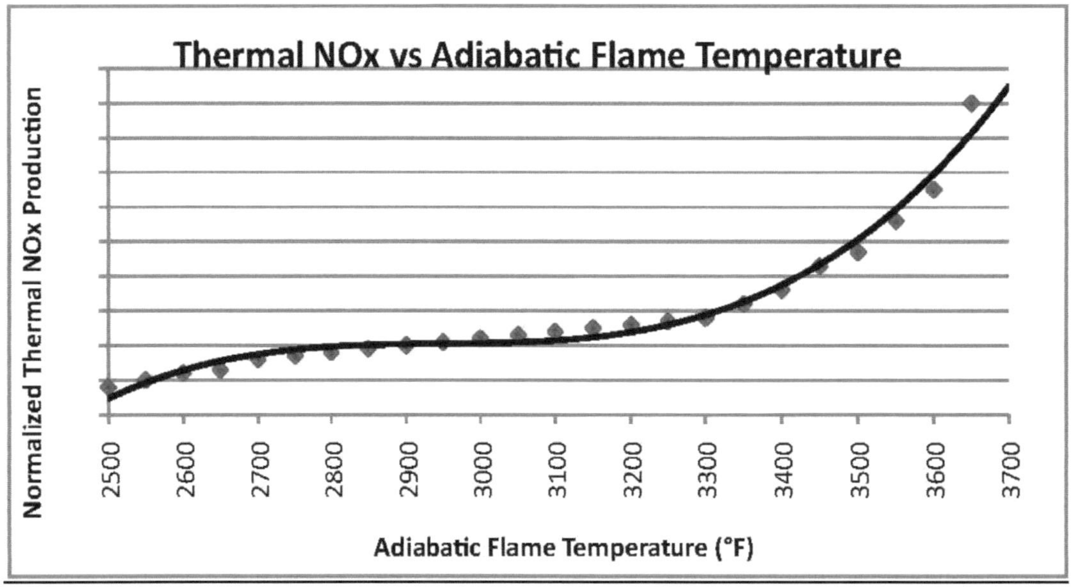

Figure 1: Calculated Peak Flame Temperature vs. Thermal NO_x Production

The industry's standard method to reduce thermal NO_x is to mix the fuel gas together with the inert products of combustion to recondition the fuel before combustion occurs. Since the reconditioned fuel is mainly comprised of inert components, the resulting composition burns at a lower peak temperature. To best utilize the inert products of combustion (flue gas) within the boiler, the fuel gas is introduced along the outside perimeter of the burner tile in an area where flue gas is present while the boiler is in operation. As the fuel gas passes through the inert products of combustion, mixing naturally occurs. This changes the composition of the fuel, and stabilization occurs at the tile exit. Since the reconditioned fuel mixture is 80 to 90% inert in most cases, the resulting flame burns at a lower peak temperature and generates less thermal NO_x.

The mixing of the fuel gas with flue gas prior to combustion is called Internal Flue Gas Recirculation (IFGR). When IFGR is too aggressive, it can result in an increased blower power usage, decreased burner turndown, and increased flame destabilization. Through Free-Jet Theory, maximizing IFGR while maintaining flame stability and flame length can become a challenge.

Flame Stabilization

In order to fully utilize the Free-Jet Theory, Zeeco devised a method to stabilize the burner flame with a highly inert fuel gas/flue gas mixture. This type of combustion is achieved when the flame is stabilized in a low-pressure area created on a series of specially designed hot refractory ledges. As combustion occurs, the refractory ledge retains heat and flame stability is enhanced. Thus, to achieve improved stability and extreme Thermal NO_x reduction, the Free-Jet Technology:

1. Mixes inert flue gas through free-jet methods with all of the fuel gas before combustion occurs, lowering flame temperature.
2. Stabilizes the flame on a refractory ledge, improving flame characteristics

Before combustion is initiated, a furnace is typically filled with normal air, which contains 21% oxygen. Once the burner is ignited, the oxygen content inside the furnace decreases until the burner achieves maximum duty. At this point, the oxygen content in the furnace is normally 2 to 3%. To keep the burner stable throughout the transition from start-up with 21% oxygen to maximum duty with 2 to 3% oxygen, Zeeco developed a series of stabilization ledges as shown in **Figure 2**. These ledges are a design feature of the Zeeco® Free-Jet Burner chosen for the boiler retrofit.

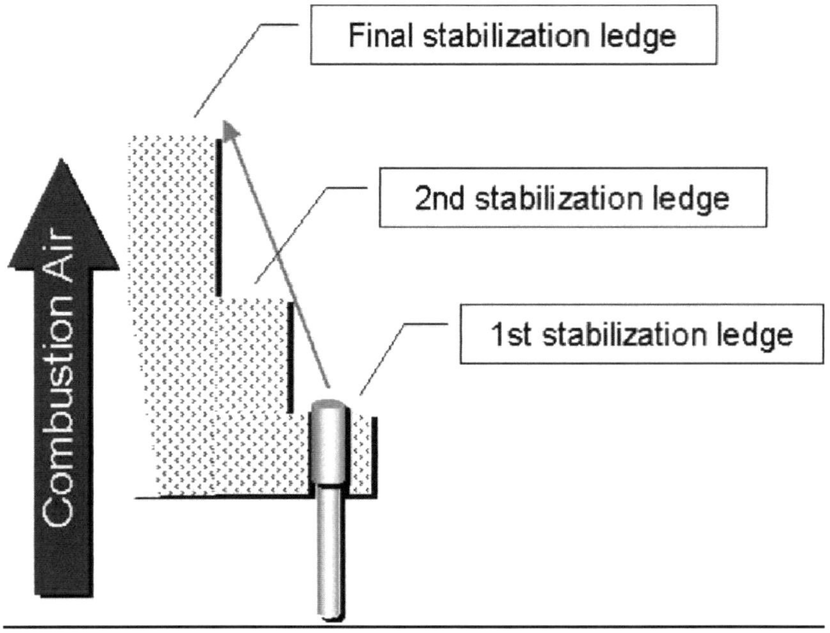

Figure 2: Free-Jet Flame Stabilization Method Illustration

Burner Technology Description

Introduction to Free-Jet Technology

The Zeeco Free-Jet burner series was designed with the specific purpose of maximizing the amount of IFGR to reduce thermal NO_x emissions without sacrificing burner performance with respect to flame length, turndown, and stability. The maximization of IFGR means many of the problems associated with using high levels of EFGR to achieve low emissions can be reduced or eliminated. Specifically, Zeeco's Free-Jet design dramatically reduces or eliminates the need for EFGR by reducing blower power usage, increasing turndown, reducing maintenance and improving flame quality.

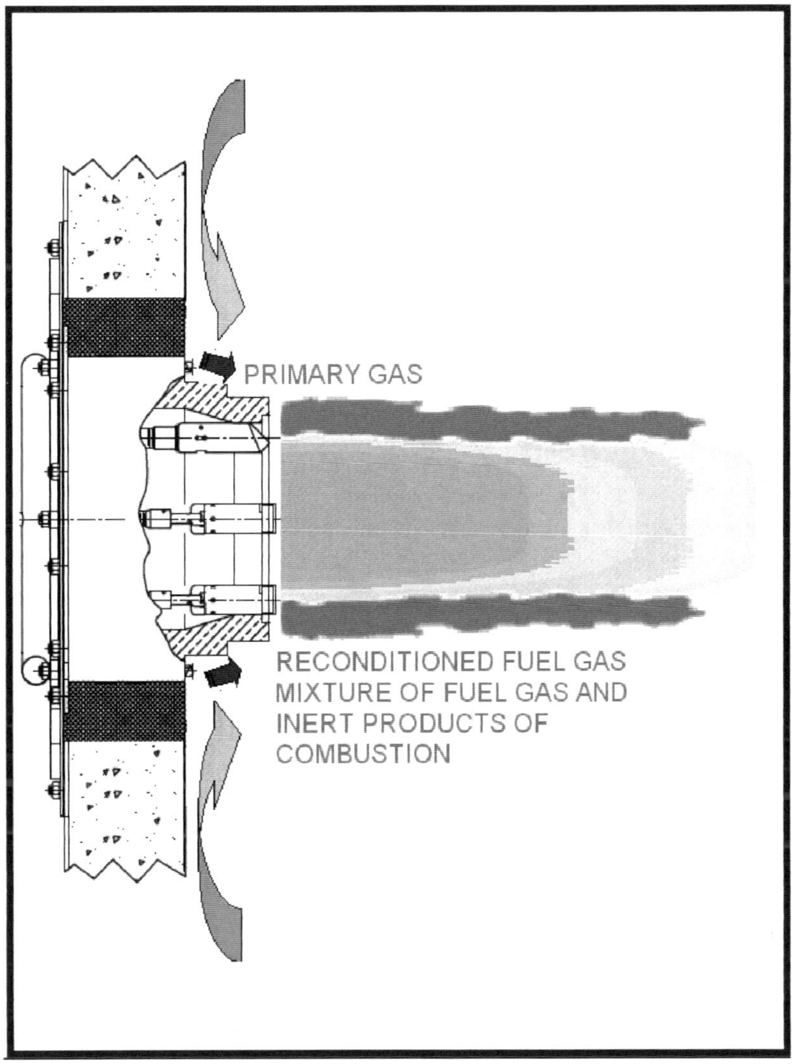

Figure 3: Inert Products of Combustion (Flue Gas)

Pre-Award Engineering

Since the boiler was only in operation for a short period of time, the plant needed assistance with the selection of the new gas firing system and options that would allow them to operate the boiler at the same steam flow capacity and pressure as originally designed for petcoke firing. Zeeco designed three possible boiler conversion options for the operator to consider. Every option required removing the existing start-up burners and closing off the tube openings or retrofitting with new larger burners:

1) Remove the fluid bed bottom and vertically fire the boiler. This option was feasible for a full 100% conversion as the entire fluid bed and nozzle/air system would be removed in favor of six new burners with a corresponding windbox and air system (1500 MMBtu/hr).
2) Removing the four start-up burners and increasing the firing capacity to the necessary levels for full load steam conditions (1500 MMBtu/hr).
3.) Removing the four start-up burners with new burners, plus adding a second level of four burners above to achieve the necessary heat input for full load steam conditions (1500 MMbtu/hr).

Option One was not selected, as it would have required an entire new lower section of the boiler, plus major reconstruction of the fluid bed in order return to solid fuel firing. Option Two was not selected, as the heat release/absorption area within the boiler would have been too great for the water/steam production side.

Option Three was selected, as the owner determined that they wanted the flexibility to return to petcoke firing should the price for natural gas rise to a level higher than petcoke fuel. This allows ~80% of the system's equipment to remain available for future petcoke firing. This solution also offered the least cost to retrofit back to solid fuel firing. The plant owner also had the challenge of the water/steam side of the project, as the boiler system would now be required to operate on 100% natural gas firing instead of a high pressure recirculating fluid bed system. Zeeco assisted the plant owner and the boiler's Original Equipment Manufacturer (OEM) with the system design and selection. Several iterations were submitted by the Boiler OEM on the redesign, and Zeeco worked as a team with the three parties to select the best option for firing the boiler.

The challenge with the boiler redesign interactions was to ensure that the following conditions were met, as required by the plant owner:

- Maintain main steam flow per original design
- NO_x emissions at permit levels
- Boiler able to follow steam needs per electrical generation requirements
- Ability to ramp per the original design conditions
- Turndown of 10-1
- No design impact or changes on existing water or steam circuit metallurgy (tube material changes)

Two design challenges were discovered throughout the Pre-Award Engineering Phase. The first included the air system redesign due to the fluid bed operating on primary and secondary air systems. This is further detailed in the Contract Execution Section below. Another challenge was the integration of the natural gas firing Burner Management System (BMS) and controls into the existing Distributed Control System (DCS). This is also further explained in the Contract Execution Section.

Project Engineering

Once awarded, Zeeco immediately began the burner and BMS systems design. The plant owner required less than 26-week delivery for all components, as the need to fire natural gas was critical to the financial status of the project.

Zeeco began the contract with a full-site visit with the entire team, including the plant owner and boiler OEM. The technical requirements and time schedule were reviewed to ensure all those involved were informed of the critical nature of the project. The technical details of the project varied, as the unit was being fully converted from a solid fuel CFB to a natural gas fired utility boiler. Many items required detailed analysis during the contract phase, including:

- Burner integration into the existing waterwall and boiler structure
- Addition of one new elevation of burners and effects of flame in this furnace region
- Complete redesign of the burner combustion air system and air heater outlet to a single pressure and delivery system to the new burners
- Integration of the new BMS panel and controls logic into the existing DCS system
- Design and location of all new gas fired fuel valving and rack systems, plus local termination cabinets
- Refractory removal and modification to support new natural gas firing (lower boiler refractory was no longer needed due to the elimination of petcoke and recirculating solids from the cyclone)
- Structural modifications to support all equipment and ancillaries for the natural gas firing (combustion air ducting, burner supports, valve racks, etc.)

Burner integration posed several challenges. The new burners were required to fit into the existing waterwall opening and modified from a downfired arrangement to a horizontal firing arrangement. Zeeco advised the plant owner that the new Zeeco Free-Jet DT burner would be able to fit within the existing opening, and would allow the exact tube opening design to be used for the new upper burner elevations. Zeeco also assisted with the design of the support systems and relocation of all structural steel for the new burner locations (maintenance platforms, etc.).

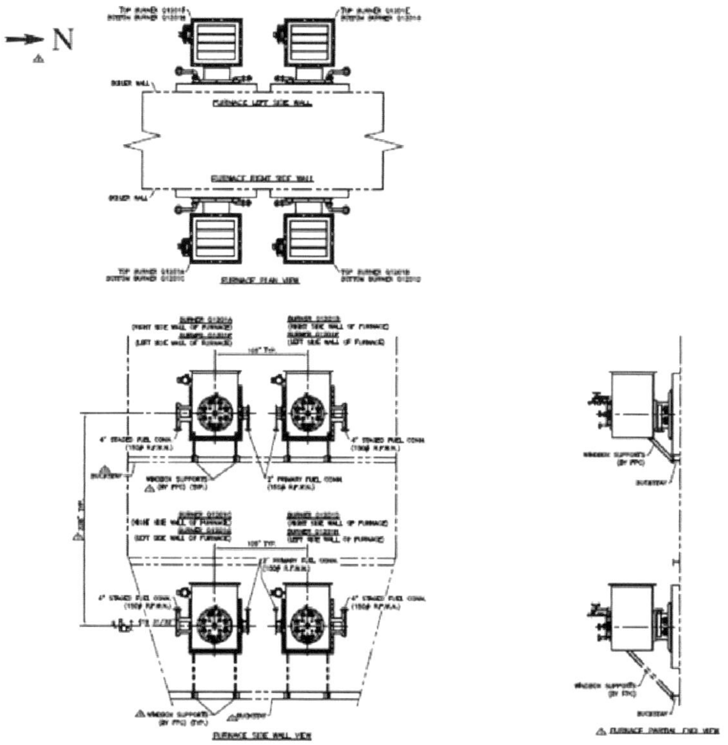

Figure 4: Boiler Design Diagram

It was important to ensure the new upper burner elevation had a proper flame fit within the boiler furnace and would not adversely affect heat transfer to the waterwalls or radiant furnace super heat. An elevation of ~19 feet was selected for the new burners above the existing burners as the ideal location, for not only proper flame fit but also ease the installation of the burners without major boiler structural modifications.

Figure 5: Zeeco's DT Free-Jet Burner

The combustion air system was completely redesigned from a two-system design (primary for the fluid bed nozzles and secondary air system for bed support), to a single pressure design. Zeeco was contracted to assist the plant owner with the entire system redesign, as well as routing to the new burner elevations. Zeeco also assisted with a new design for the air heater outlet and fan control due to the single pressure design needed for the new burners. Zeeco's Model Simulation Engineering Team spent significant time on-site with the Plant Owner's personnel designing the new system. Zeeco's engineers designed and constructed a scaled Plexiglas model of the entire combustion system and ran the Physical Flow Simulation model in our Connecticut facility. The completed model ensured +/- 3% distribution to each burner with less than 10% peripheral distribution around each burner's air entry.

Figure 6: Free-Jet Physical Flow Simulation

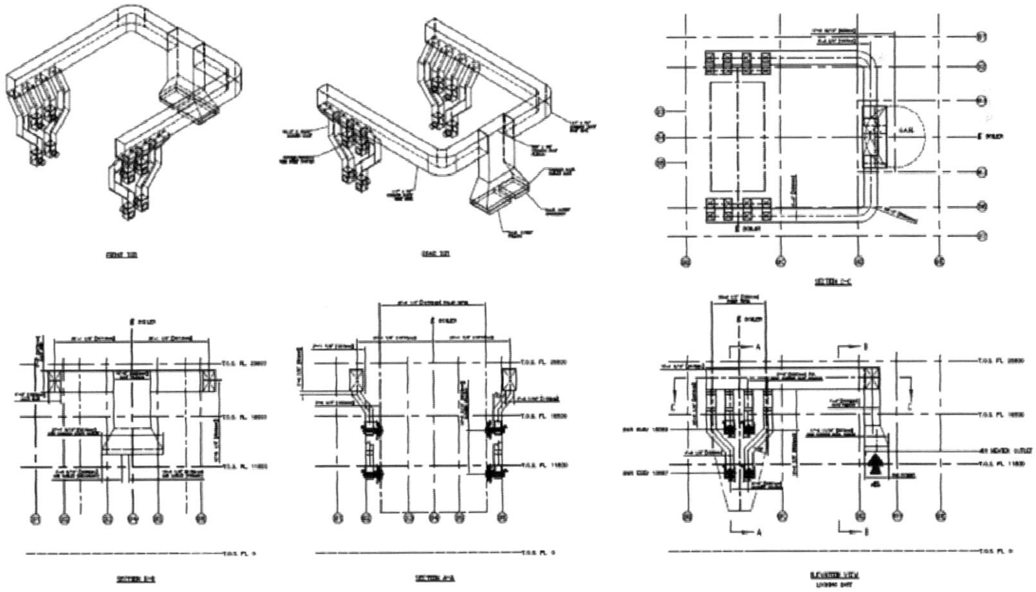

Figure 7: Zeeco's New Design Combustion Air System

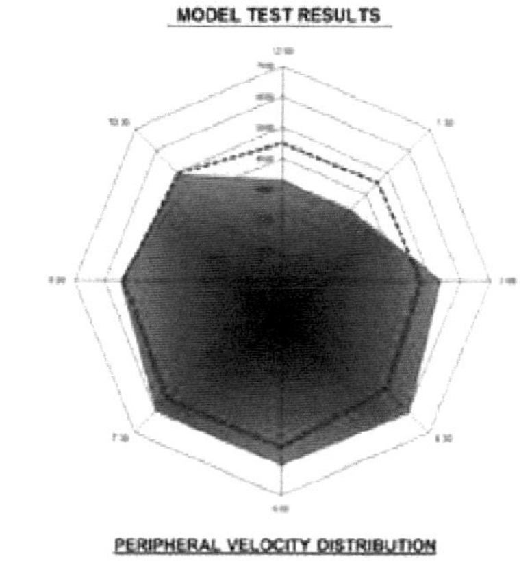

Figure 8: Example of Model Study Peripheral Distribution Chart

Due to the boiler's original controls and Burner Management System's design for main fuel petcoke firing and natural gas start up burners, they were not designed for full load natural gas firing. During the initial design proposal phase, it was agreed that a separate Redundant PLC BMS system would be necessary to integrate into the main DCS system. This would allow the main system to remain "as-is" in the event the unit was to return to petcoke firing in the future. Essentially, it would be a "fuel selector" switch to return the boiler's operation back to petcoke firing. This system was designed in working with the plant owner's original DCS and BMS supplier to ensure a smooth integration into the DCS system.

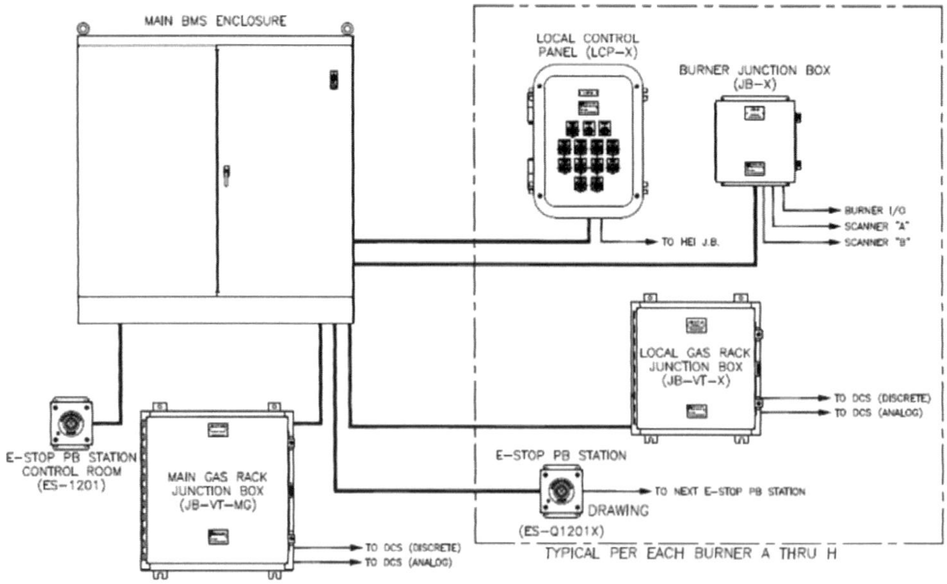

Figure 9: Boiler's Burner Management System

In order for the new burner system to be modified to fire on natural gas, several modifications were required for successful full load natural gas firing, including:

- Removal of furnace's lower slope refractory- During petcoke firing, the turbulent nature of the fluid bed, including the solids return from the cyclone, required the lower slope tubing to be fully protected from potential erosion. When the solid fuel is removed, this lower slope no longer required the refractory coverage to protect the tubing from solids within the cyclone, promoting further heat absorption in the furnace for natural gas firing, and to assist with lower emissions.

- As noted above, the cyclone design in a CFB is to remove solids from the furnace exit flue gas and return them to the fluid bed to promote high efficiency firing and lower carbon loss. With petcoke firing removed, the cyclone was no longer needed for this operation. The inlet to the cyclone plated off, effectively "closing" the inlet so no flue gas would pass through the cyclone. The loop seal exit was "plugged" with refractory to protect it from furnace gasses. This sealing can be deemed 'temporary", as the plate and refractory plug can be removed in the event petcoke firing is deemed cost effective over natural gas firing.

- All back-end clean up equipment are placed out of service due to natural gas firing.

Upon all final design "freeze" and plant owner approvals, fabrication and procurement proceeded and all equipment was delivered to the site by the end of March 2013. Due to the critical schedule, the plant owner was required to take the boiler out of service in late January 2013 to begin the modification process for the natural gas firing. The equipment, upon on-site arrival, was immediately transported to the boiler and the installation process began.

Due to the diligence of the entire engineering team, the integration and installation of the burners, valving and BMS proceeded very smoothly throughout the outage. All the upfront engineering and time ensured the design was cost effective and the outage was completed on time.

Commissioning and Startup

The plant owner also contracted Zeeco's Service Engineers to provide full commissioning and installation assistance, to ensure that all mechanical and electrical equipment was installed per Zeeco's requirements. Zeeco's Service Engineering team was on-site to review and check out the electrical/mechanical/BMS/DCS installation. This proceeded very smoothly during checkout, and all minor issues were addressed in the time allotted for commissioning. Actual equipment start-up proceeded immediately after the final commissioning. Burner performance and full steam boiler operation was achieved in less than two weeks.

Final Conditions are as noted:

- Final emissions performance met the contract and permit requirements of 0.07lbs/MMBtu
- Full load steam boiler requirements were met within the two week timeframe

Conclusion

Boiler steam load and emissions were met in a very timely fashion, and the plant owner was able to generate plant power within less than a month after the initial boiler start-up. The plant owner operated the boiler and turbine/generator near design capacity for over three months after final commissioning. The effort placed on the entire project, pre and post-award, led to the overall success of the boiler's conversion. The integration by the plant owner and the three primary engineering parties led to a fully integrated design that was reviewed by all to ensure the timely success of the project. Based upon the success of the first boiler, Zeeco was awarded the contract for the second CFB boiler conversion, and will begin with installation and commissioning to be in mid-2014.

351797 Vapor Intrusion - Models and Their Value

Abstract:

Vapor Intrusion- Models and Their Value

The problem of vapor intrusion into structures built atop contaminated properties has received increasing attention over the last decade. On the national level, new guidance from the US EPA is imminent, but in the meantime different states have adopted a wide range of policies to address the problem. Many approaches to characterization are empirical, based upon measurements at impacted locations. Since these can involve alarming and intrusive measures involving many residents, some emphasis has been placed on screening, using simple mathematical descriptions of the problem. With the vastly increased availability of powerful fluid dynamics codes, these models can guide more of the site investigation work, as well as evaluate possible possible mitigation strategies. Examples will be given of how such tools can be used to assess both petroleum-based and chlorinated solvent-based situations, based upon recent research carried out in Brown University's Superfund Research Program.

The approach involves solving the equations of vapor transport in the subsurface below the structure of interest, which is typically the soil of the vadose zone. The transport is dominated by diffusion, though entry into the structure may be strongly influenced by local advective transport near the building foundation, as a result of small pressure differences between ambient and the indoors. The indoor environment is typically modeled using a CSTR model. Subtle factors can have a very large influence on predicted indoor air exposures. These include effects related to the resistance of a capillary zone in the case of a chlorinated solvent source in the groundwater, or the form of the assumed biodegradation rate law in the case of petroleum contaminants. The models point to what might represent reliable or problematic field measurement strategies. Also, once a reliable site model is developed, the efficacy of mitigation methods such as subslab depressurization, may be more reliably examined

(137c) Synthesis, Fine Structural Characterization, and CO2 Adsorption Capacity of Cobalt and Nickel-Based Metal Organic Framework-74

Adhikari, A. K., Yuan Ze University

Lin, K. S.,

Originally presented on: 4/2/2014 14:30:00 - 15:00:00

Abstract: Two metal organic frameworks of MOF-74 group (cobalt and nickel-based) were synthesized, characterized, and evaluated for CO_2 adsorption. The BET specific surface areas of MOF-74(Co) and MOF-74(Ni) MOFs were 1,406 and 1,418 m^2g^{-1}, respectively with the identical pore volume of 0.84 cm^3g^{-1}. CO_2 adsorption capacities of MOF-74(Co) and MOF-74(Ni) were 3.35 and 4.07 $mmol{\cdot}g^{-1}$, respectively measured at 273 K and 1.1 bar. The oxidation state of central atoms in MOF-74(Co) was composed of Co(II) and Co(III) confirmed by XANES spectra while MOF-74(Ni) was Ni(II) central atoms. The bond distances of Co??O and Ni??O were 1.95 and 1.97 Å, respectively.

1. Introduction

Global population and the industrialization are increasing and hence, the energy consumption is also increasing swiftly. Till now, the main source of energy is the burning of fossil fuels. However, burning of fossil fuels releases a vast amount of CO_2 greenhouse gas into atmosphere which causes the raise of average surface temperature of the earth. As the present developments in CCS technology are not sufficiently effective by means of cost and energy consumption, it has prompted to improve the existing technologies and find some new alternatives [1]. Metal organic frameworks (MOFs) are a new class of materials that have attracted substantial attention in past few years [2]. MOFs are synthesized using organic linkers and metal joints that self-assemble to form different dimensional networks. It can generate unusually large diameter channels and cavities which is particularly attractive for CO_2 gas storage application. Here we report the CO_2 gas adsorption capacity on same type of two MOFs; such as MOF-74(Co) and MOF-74(Ni).

2. Experimental

In a typical experiment, 0.75 g of 2,5-dihydroxyterephthalic acid (DHTA) and 3.0 g of $Co(NO_3)_2{\cdot}6H_2O$ or $Ni(NO_3)_2{\cdot}6H_2O$ were dissolved in 150 mL of DMF with sonication in a 250 mL Duran bottle. The bottle was capped tightly and placed in a 110°C oven for 20 h. After cooling to room temperature, the mother liquor was decanted and the crystals were washed with DMF. The product was immerged into methanol for six days. The products were then evacuated and heated under vacuum to150°C over a period of 12 h.

3. Results and discussion

The FE-SEM images of MOF-74(Co) and MOF-74(Ni) samples are shown in Fig.1a and 1b, respectively. It shows that MOF-74(Co) has cauliflower type structure with the particle size of about 4-6 mm and MOF-74(Ni) has uniformly distributed spherical shape with the average diameter of 2-4 mm, although some particles are aggregated. The XRD patterns (Fig.1c) shows that both samples have the unique crystalline phase indexed on the basis of a face-centered hexagonal unit cell of MOF-74. Both samples indicate almost similar diffraction patterns at identical 2?Æ values ascribed the formation of similar crystallographic phases.

Figure 1. -SEM images of (a) MOF-74(Co), (b) MOF-74(Ni), and (c) XRD patterns of synthesized MOF-74 samples.

MOF-74(Co) was thermally stable up to 258°C while in case of MOF-74(Ni) the value was 344°C confirmed from the TGA analysis shown in Fig. 2. The thermogravimetric curves of both samples are consistent with prominent weight-loss steps. It can be seen that the thermal stability is not similar for these two samples although they follow the same trend of loosing weight. In case of MOF-74(Co), a fully hydrated molecular sieve contains about 27 wt.% of water and other guest molecules. and in case of MOF-74(Ni) this value is almost similar, but there is one additional step of weight loss of about 13 wt%.

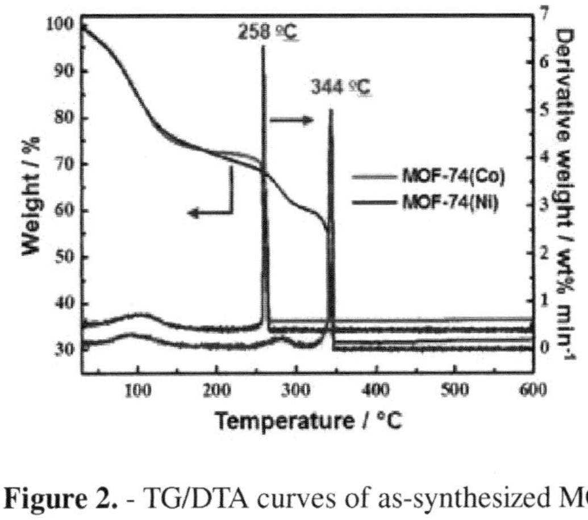

Figure 2. - TG/DTA curves of as-synthesized MOF-74(Co) and MOF-74(Ni)

From the N_2 adsorption isotherms it has been evaluated that both sample follow type-I isotherms with almost no hysteresis loop which corresponding to microporous material. The BET specific surface areas of MOF-74(Co) and MOF-74(Ni) MOFs were 1,406 and 1,418 m^2g^{-1}, respectively with the identical pore volume of 0.84 cm^3g^{-1}. DFT pore size distribution of MOF-74(Zn) and MOF-74(Cu) are shown in Fig. 4a and Fig. 4b, respectively while the values of average pore size and volume are presented in Table 1. These results suggest that these samples are microporous materials.

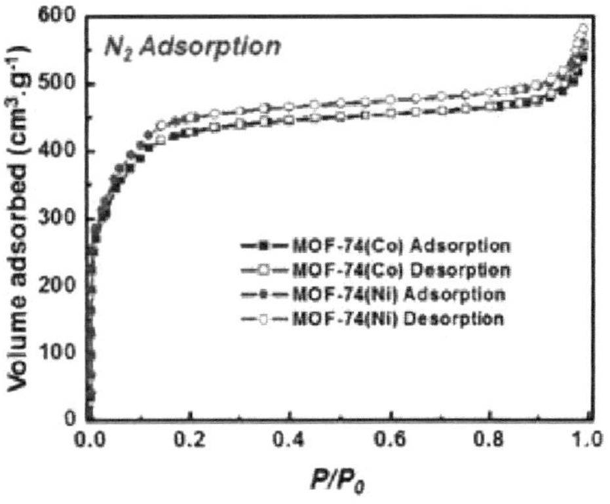

Figure 3. - N_2 adsorption/desorption isotherms of as-synthesized MOF-74(Co) and MOF-74(Ni). Filled and open symbols represent the adsorption and desorption of nitrogen gases, respectively.

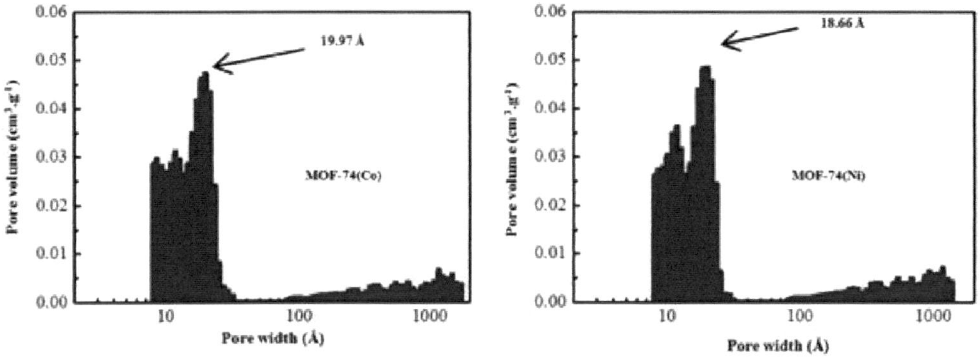

Figure 4. DFT pore size distribution of (a) MOF-74(Co) and (b) MOF-74(Ni)

Table 1. Fine structural parameters of MOF-74 samples analyzed by Co and Ni K-edge EXAFS.

Sample	BET SSA (m^2g^{-1})[a]	Langmuir SSA (m^2 g^{-1})	Pore volume (cm^3g^{-1})[b]
MOF-74(Co)	1,406	2,120	0.87
MOF-74(Ni)	1,418	2,123	0.84

[a] *The specific surface area was measured by BET method; and*
[b] *Pore volume was measured by density functional theory (DFT) method.*

Fine structure and oxidation state of Zn and Cu species in MOF-74 were further investigated through X-ray absorption spectroscopy. The data were collected several times and standard deviation was also calculated. The XANES region of the X-ray absorption spectrum is very informative on both the oxidation and coordination states of metal ions. XANES spectra of MOF-74(Co) and MOF-74(Ni) samples are shown in Fig. 5a and 5b, respectively along with the standard materials. The comparison is designed to determine the oxidation states of center atoms in MOFs and understanding the effects of DHTA linkers and guest molecules on the crystal structures. It was found that the edge position (7709.46 eV) of as-synthesized MOF-74(Co) sample is very close to the Co(II). However, compared with the standard Co(II) the offset was +0.46 eV, revealed that Co ions were not entirely surrounded by oxygen atoms and hence formed hydroxides (OH^-) through covalent bond. This result reveals that the as-synthesized sample was not totally moisture or guest molecule free. The source of the

OH⁻ ions is either adsorbed moisture or the organic linkers that cause the little distortion in the crystal structure. In case of MOF-74(Ni) the offset was +0.57 eV. Fine structural parameters for both samples are shown in Table 2 where it can be seen that the bond distances of Co??O and Ni??O were 1.95 and 1.97Å, with the coordination number of 4.27 and 4.23, respectively.

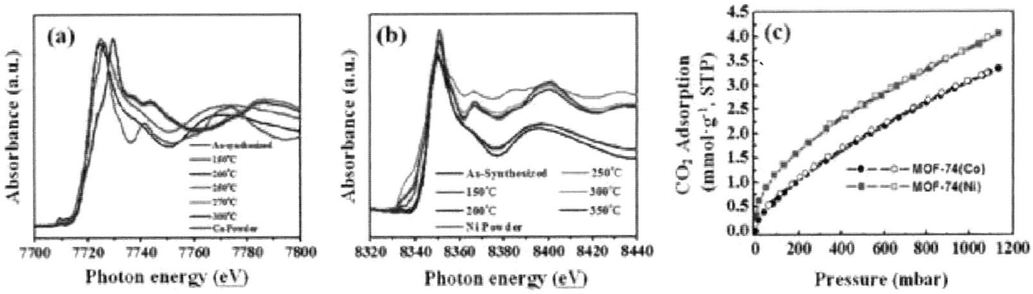

Figure 5. (a) Co K-edge XANES spectra of MOF-74(Co) samples with Co(0) standard, (b) Ni K-edge XANES spectra of MOF-74(Ni) with Ni(0) standard, and (c) CO_2 adsorption isotherms of MOF-74 samples measured at 273 K. Filled and open circles represent the adsorption and desorption, respectively.

Table 2. Fine structural parameters of MOF-74 samples analyzed by Co and Ni K-edge EXAFS.

Sample	Shell	$CN^a \pm 0.5$	$R^b \pm 0.002$ (Å)	$\Delta\sigma^{2c}$ (Å²)
MOF-74(Co)	Co—O	4.27	1.953	0.0037
MOF-74(Ni)	Ni—O	4.23	1.974	0.0038

[a]*Coordination number;* [b]*Bond distance;* [c]*Debye-Waller factor.*

The CO_2 adsorption capacities of the synthesized samples are shown in Fig. 5c. The steady ascending slope of the CO_2 adsorption isotherms represents strong adsorption of CO_2 molecules having quadrupoles onto the open metal sites in MOF-74 samples. The CO_2 adsorption capacity of MOF-74(Ni) measured at 273 K and 1.1 bar was 4.07 mmol·g^{-1}; significantly higher than MOF-74(Co) which was 3.35 mmol·g^{-1}. However, both results are much higher than other prospective adsorbents of CO_2 such as Zeolite 13X [3] and ZIF-96 [4]. The elevation in CO_2 adsorption in MOF-74(Ni) may attribute to the higher surface area than MOF-74(Cu) and more importantly the strong affinity to CO_2 molecules due to different electronic configurations. It can be seen that the adsorption and desorption cycles are almost reversible for both samples. This integrity was also corroborated by the N_2 adsorption. The steady uphill slope of the adsorption curves indicates that higher adsorption capacity may obtain at higher pressure.

4. Conclusions

The MOF-74 samples were capable to store significant amount of CO_2 at relatively lower pressure. The difference in CO_2 adsorbed amount between two samples may attribute to the different affinity of cobalt and nickel ions to CO_2 molecules as well as the specific surface area. The moderate thermal stability confirms that these MOFs may apply to control the CO_2 gas from the emission of refinery or other industries.

References

1. A. Torvanger, M. T. Lund, N. Rive, Mitig. Adapt. Strat. Gl. 18 (2013) 187.

2. T. G. Glover, G. W. Peterson, B. J. Schindler, D. Britt, O. M. Yaghi, Chem. Eng. Sci. 66 (2011) 163.

3. J.-S. Lee, J.-H. Kim, J.-T. Kim, J.-K. Suh, J.-M. Lee, C.-H. Lee, J. Chem. Eng. Data 47 (2002) 1237.

4. W. Morris, B. Leung, H. Furukawa, O. K. Yaghi, N. He, H. Hayashi, Y. Houndonougbo, M. Asta, B. B. Laird, O. M. Yaghi, J. Am. Chem. Soc. 132 (2010) 11006.

Treatment of Flowback Water from Hydraulic Fracturing with Biochar

Maoqi (Mark) Feng
Steven Cooks
Zhigang Feng

April 2, 2014

AIChE 2014 Spring Conference, New Orleans

Chemical Engineering Dept., Div. of Chemistry and Chemical Engineering, Southwest Research Institute, San Antonio, TX 78238

Outline

- Introduction
- Technical Approach
- Experimental Work
- Summary & Conclusions

2

Water Demand in Hydraulic Fracturing

- Water demand is huge: 100 billion gallons of water used for 39,000 wells in 2011
- Fracking accounts for 50% of total water consumption in some Texas areas

3

Water Requirement in Shale Gas Production

- Water requirements for hydraulic fracturing, by shale plays in Texas

Water Requirements (gallons per well)	Low Value	Median Value	High Value
Barnett Shale	<1 MM	2.6 MM	>8 MM
Haynesville and Bossier Shale	< 1 MM	5.5-6 MM	>10 MM
Eagle Ford Shale	1 MM	6-6.5 MM	13 MM
Woodford, Pearsall, and Barnett-PB Shale	<1 MM	0.75-1 MM	<5 MM

4

Flowback Water Recycling

- Half of fracturing wells are in water-stressed areas
- Flowback water reuse is needed

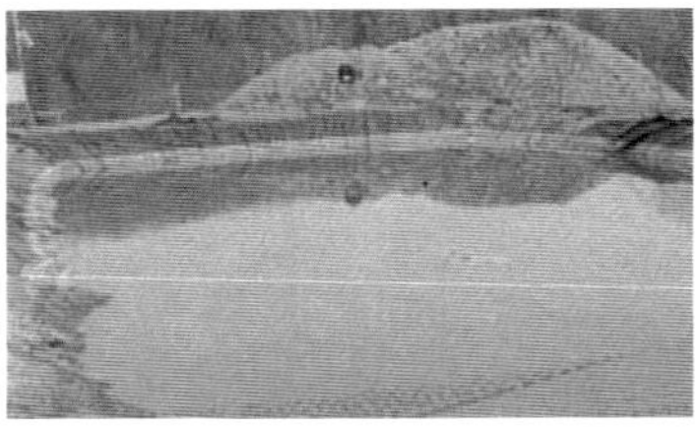

5

Reusable Water Quality

- Parameter Range

TDS at 180°C, mg/l	9,000–16,000
Turbidity, NTU	0-5
pH	6.5-8
Iron, mg/l	1-10
Chloride, mg/l	5,000-10,000
Potassium, mg/l	100-500
Calcium, mg/l	50-250
Magnesium, mg/l	10-100
Sodium, mg/l	2,000-5,000
Boron, mg/l	0-20

6

Flowback Water Treatment

- Membrane method is expensive and cannot handle solids and hydrocarbons
- Zeolites generate more solid waste
- Bioremediation is not efficient
- Nanofiltration is easily plugged
- Activated carbon adsorption is more expensive

7

Why Biochar

- Biochar is a good sorbent
- Green
- Biodegradable
- Soil compatible and amendment
- Easy to prodcue
- cost effective
- Mass production available

8

Biochar vs. Activated Carbon

- Biochar is produced at lower temperature
- Biochar is a precursor for activated carbon
- Activated carbon needs an activation step
- Biochar is more feedstock flexible

9

Biochar Production

- Biochar production from wood chips by pyrolysis
- Temperature 350°C – 550°C

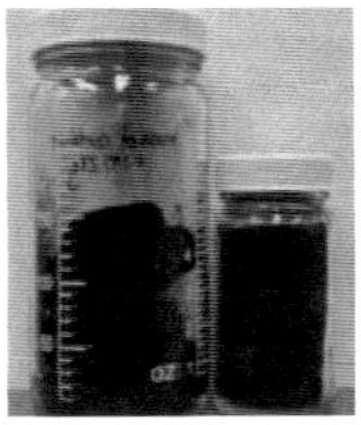

10

Biochar Property

- Pine wood biochar carbon content vs. pyrolysis temperature

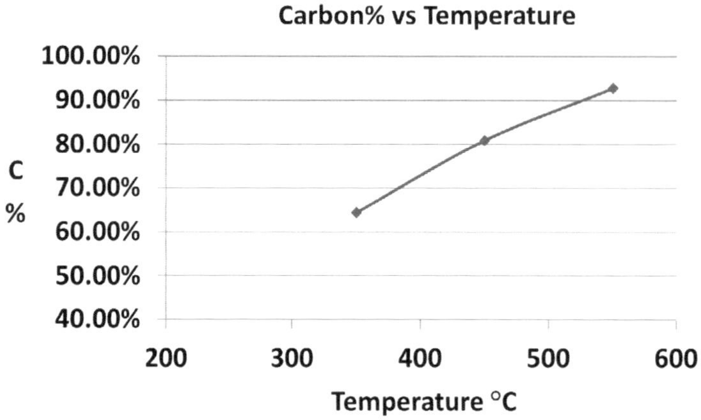

11

Biochar Property

- Oak wood biochar carbon content vs. pyrolysis temperature

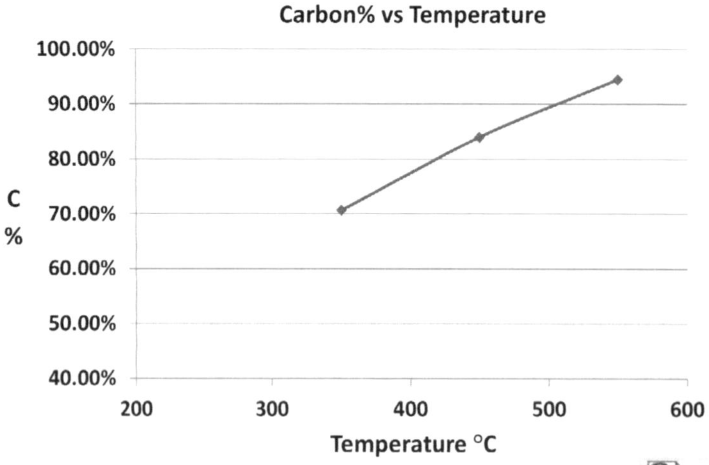

12

Biochar Preparation

- Biochar prepared from different sources at 350°C

Feed	Carbon%
Cellulose	90.09%
Pine Wood	66.43%
Oat Hulls	58.59%

13

Batch Adsorption

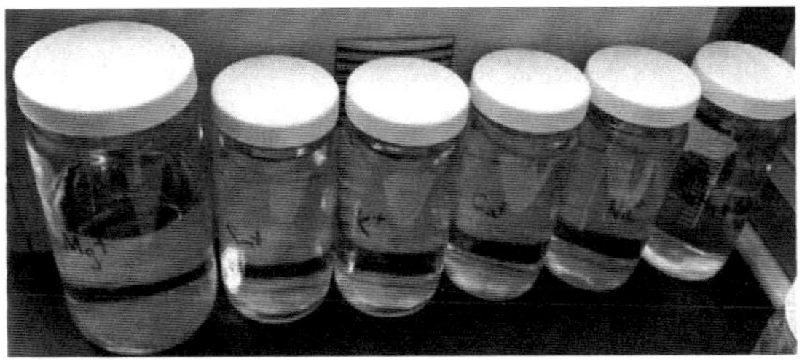

14

Batch Adsorption

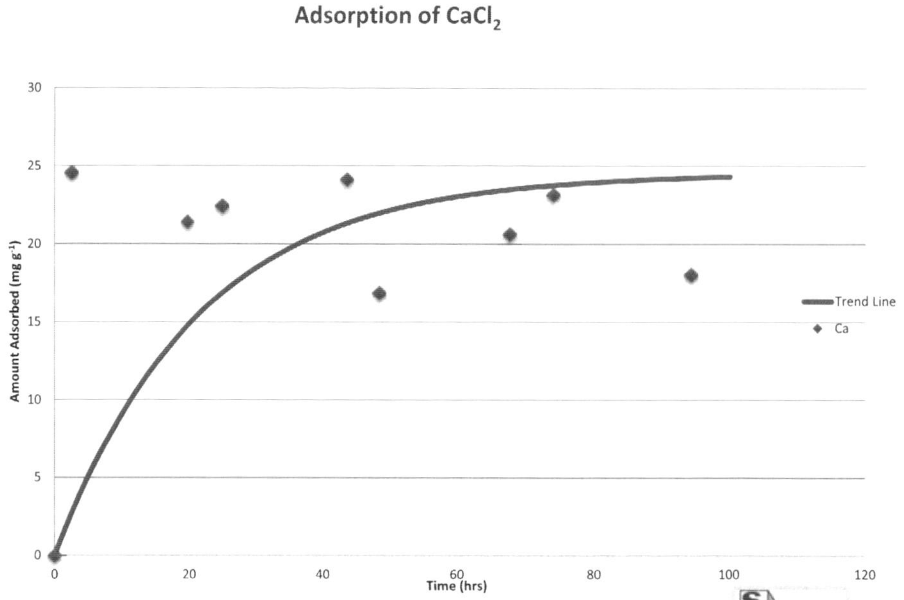

15

Sorbent Bed Evaluation in Continuous Flow Mode

- Quick evaluation of sorbent capacity in continuous flow mode – more efficient than batch mode

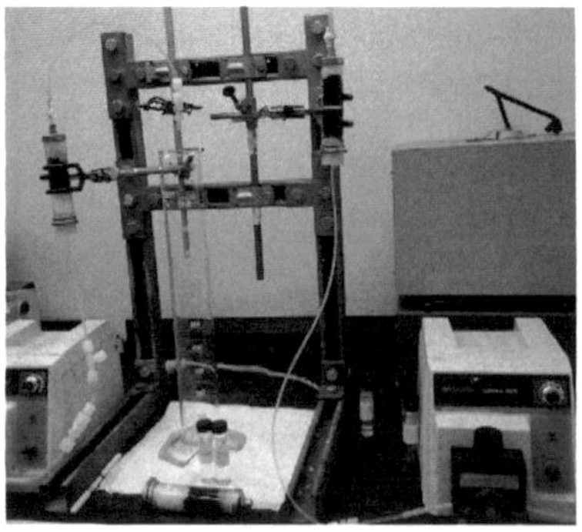

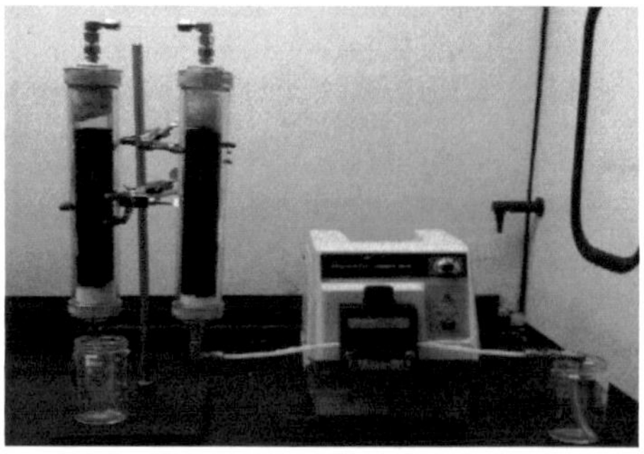

16

NaCl Removal

- NaCl removal

Time (hrs)	NaCl Mass (g)	Water Mass (g)	NaCl Concentration, wt%	NaCl wt% Adsorbed
0.00	12.75	247.5	4.90%	0.00%
2.62	0.8921	18.5788	4.58%	0.32%
19.53	0.9598	19.3488	4.73%	0.17%
24.57	0.9886	19.7544	4.77%	0.13%
43.37	0.9908	20.1408	4.69%	0.21%
48.32	0.9065	18.6041	4.65%	0.25%

17

Mg Removal

- $MgCl_2$ removal

Time (hrs)	$MgCl_2$ Mass (g)	Water Mass (g)	$MgCl_2$ Concentration, wt%	$MgCl_2$ wt% Adsorbed
0.00	4.8754	38.5514	11.23%	0.00%
0.5	4.1715	38.4528	9.79%	1.44%
1.0	4.3688	39.0205	10.07%	1.16%
1.5	4.6866	39.086	10.71%	0.52%
2.0	4.3229	38.8608	10.01%	1.22%

18

Ca Removal

- $CaCl_2$ removal

Time (hrs)	$CaCl_2$ Mass (g)	Water Mass (g)	$CaCl_2$ Concentration, wt%	$CaCl_2$ wt% Adsorbed
0.00	3.975	38.7583	9.30%	0.00%
0.5	3.9614	39.2739	9.16%	0.14%
1.0	3.936	38.9826	9.17%	0.13%
1.5	3.9684	39.5214	9.12%	0.18%
2.0	3.9067	39.0601	9.09%	0.21%
2.5	3.9021	39.369	9.02%	0.28%

19

Sr Removal

- $Sr(NO_3)_2$ removal

Time (hrs)	$Sr(NO_3)_2$ Mass (g)	Water Mass (g)	$Sr(NO_3)_2$ Concentration, wt%	$Sr(NO_3)_2$ wt% Adsorbed
0.0	0.4834	20.7043	2.28%	0.00%
0.5	0.4557	20.665	2.16%	0.12%
1.0	0.4491	21.1634	2.08%	0.20%
1.5	0.4191	20.3263	2.02%	0.26%
2.0	0.4114	20.5186	1.97%	0.32%
2.5	0.4031	20.1342	1.96%	0.32%

20

Inorganics Removal

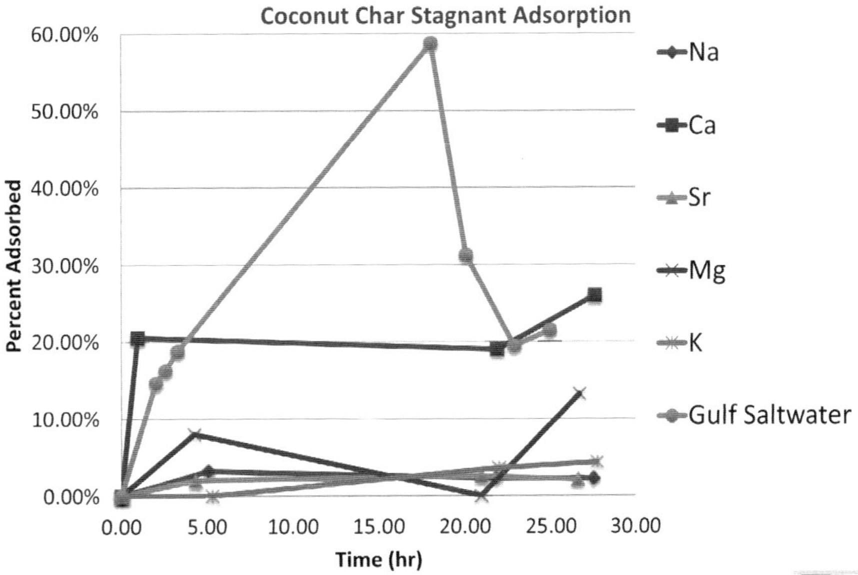

21

Hydrocarbons Removal

- Removal of diesel hydrocarbons in water by biochar adsorption
- Per pass removal 87%
- Higher adsorption capacity for biochar produced at higher temperature

22

Solid Particulates Removal

- Removal of solid particulates in flowback is necessary
- Pump damage
- Long settling time
- Fine solids removal is still in progress

23

Summary

- Biochar production from wood chips
- Quick evaluation of sorbent capacity in continuous flow mode
- Successful removal of inorganic salts: NaCl, K^+, $MgCl_2$, $CaCl_2$, Sr^{2+}
- Successful inorganics removal from sea water
- Successful removal of hydrocarbons from simulated flowback water

24

Some Thoughts

- Biochar adsorption bed can be used with other methods, such as ultrafiltration, reverse osmosis, and ion exchange units

- Biochar sorbent bed can be used as a pretreatment or post-polishing step for other methods

25

Acknowledgement

Thank you!

Questions?

26

352927 Hazards Management for Building Locations through Air Dispersion Modeling

Abstract:

Oil, gas, and petrochemical industries have reached a critical era involving its timeworn infrastructure. Previously, facilities degraded with minimal effort allocated for updates that ensure personnel safety (i.e., process inventory isolation, automative control systems, etc.). More recently, industry as a whole is progressively adopting a more safety conscientious attitude about personnel and the environment. These personnel and environmental concerns are driving new codes, standards, recommended practices, and regulations to be developed and implemented. Safety studies are being performed which assess facilities to determine if these new regulations are met, while providing key insight to eliminate any deficiencies that are discovered. Available advanced air dispersion models estimate or predict the concentration of air pollutants or toxicants emitted from various sources such as industrial plants, road traffic vehicles or accidental chemical releases. Facility siting assessments are amongst the required studies used for managing the risk to on-site personnel located in new and existing buildings intended for occupancy, where air dispersion models can be used to determine the consequences of accidental releases of hazardous or toxic materials. This paper discusses case studies carried out for hazards management for building locations with different air dispersion modeling software tools including: SCREEN3, AERMOD, ALOHA, and PHAST.

353139 H2S Removal from Biogas Using an Ionic Liquid as Physical Absorbent

Abstract:

Biogas is nowadays a valuable renewable energy carrier which can be used directly as a fuel or as a raw material for the production of synthesis gas and/or hydrogen. Methane (CH_4) and carbon dioxide (CO_2) are the main constituents, but biogases also contain significant quantities of undesirable compounds (contaminants), such as hydrogen sulfide (H_2S) and ammonia (NH_3). The existence and quantities of these contaminants depend on the biogas source (i.e., landfills, anaerobic fermentation of manure). Their presence constitutes a major problem because (1) they can adversaly affect any biogas thermal or thermocatalytic conversion device (e.g., corrosion, erosion, fouling); and (2) they produce harmful environmental emissions. It is therefore important to develop a reliable and efficient biogas purification process. This work is aimed at presenting an alternative removal approach for H_2S from biogas based on physical absorption using an ionic liquid as solvent.

AUTHOR INDEX

AUTHOR INDEX